Geology of the Lower Gila Region of Arizona

by US Dept of Interior

with an introduction by Kerby Jackson

Introduction

It has been decades since the Department of Interior released their important publication "Geology of the Lower Gila Region, Arizona". First released in 1921, this important volume has been out of print and has been unavailable to the mining community since those days, with the exception of expensive original collector's copies and poorly produced digital editions.

It has often been said that "*gold is where you find it*", but even beginning prospectors understand that their chances for finding something of value in the earth or in the streams of the Golden West are dramatically increased by going back to those places where gold and other minerals were once mined by our forerunners. Despite this, much of the contemporary information on local mining history that is currently available is mostly a result of mere local folklore and persistent rumors of major strikes, the details and facts of which, have long been distorted. Long gone are the old timers and with them, the days of first hand knowledge of the mines of the area and how they operated. Also long gone are most of their notes, their assay reports, their mine maps and personal scrapbooks, along with most of the surveys and reports that were performed for them by private and government geologists. Even published books such as this one are often retired to the local landfill or backyard burn pile by the descendents of those old timers and disappear at an alarming rate. Despite the fact that we live in the so-called "Information Age" where information is supposedly only the push of a button on a keyboard away, true insight into mining properties remains illusive and hard to come by, even to those of us who seek out this sort of information as if our lives depend upon it. Without this type of information readily available to the average independent miner, there is little hope that our metal mining industry will ever recover.

This important volume and others like it, are being presented in their entirety again, in the hope that the average prospector will no longer stumble through the overgrown hills and the tailing strewn creeks without being well informed enough to have a chance to succeed at his ventures.

Kerby Jackson
Josephine County, Oregon
May 2016

GEOLOGY OF THE LOWER GILA REGION, ARIZONA.

By Clyde P. Ross.

INTRODUCTION.

In 1917 and the early part of 1918 the writer made an investigation of desert watering places and routes of travel in a part of southwestern Arizona. The results of this work are to be published in two water-supply papers of the United States Geological Survey—a preliminary report giving information in regard to roads and watering places and a final report which is to include also much miscellaneous information on the geology, geography, and hydrology of the region. In the present report the geologic information obtained in the course of the work is summarized. As the geologic investigation was necessarily of a reconnaissance character, the information obtained is incomplete, but much of it is new and it is hoped will prove of value.

The area covered lies in the central part of Yuma County and the western part of Maricopa County, Ariz. In Maricopa County it includes an irregularly triangular region with Phoenix at its eastern vertex, bounded on the north and northeast by the road from Phoenix through Wickenburg to Wenden and on the south by the valleys of Salt and Gila rivers and extending westward to the county boundary. In Yuma County the area is bounded on the north by the road between Wenden and Parker through Cunningham Pass and on the south by the valley of Gila River and extends entirely across the county to the western boundary, Colorado River.

The commercial development of such a region as the one here described is intimately related to the geology. The hope of finding mineral deposits usually furnishes one of the initial incentives for pioneering in such regions. When promising deposits are found, as they have been here, towns spring into existence and the settlement of the country commences.

In the early days in southwestern Arizona fur trapping vied with prospecting as an occupation for the adventurous frontiersmen. When the country became a little better known and more settled, cattle raising and farming were introduced. Both of these industries, particularly farming, depend for their success on a supply of water. The available surface water here soon proved insufficient, and the settlers began to utilize the ground water by means of wells. The distribution, quantity, and quality of the ground water in a region are directly dependent on the geology and physiography of the region.

ROCK FORMATIONS.

At first glance most of the mountains in this section of the country present a very similar appearance. Examination soon shows, however, that they are composed of rocks of a number of very diverse types. There are great masses of ancient metamorphic rocks, of granites and granitic gneisses, and of lavas and tuffs belonging to at least two distinct periods, together with subordinate amounts of sediments associated with the older lavas and tuffs, and sand and gravel filling the valleys between the ranges. More detailed work will undoubtedly result in still further subdivision of the rocks. The metamorphic rocks certainly represent two and probably more than two periods. The granitic rocks belong to at least two periods of intrusion.

BASAL COMPLEX.

Definition.—Highly metamorphosed sedimentary rocks with associated granitoid gneisses and other rocks of igneous origin make up the whole or a large part of many of the mountain ranges in this region. These rocks will be referred to collectively as the basal complex. They may be divided into four general groups—(1) igneous rocks, (2) highly metamorphosed schistose rocks, probably in the main of sedimentary origin, (3) thoroughly metamorphosed but much less schistose sedimentary rocks separated from those of the second group by an unconformity, and (4) metamorphosed but not schistose limestone and quartzite, the youngest sedimentary rocks in the basal complex. The igneous rocks may be further subdivided into batholithic masses with associated dikes and a group of somewhat younger dikes which cut the less metamorphosed portions of the basal complex.

Distribution and character.—This ancient complex is present in every mountain range and almost every range of hills in the region under consideration. Even in those mountainous areas where it is not shown on the geologic map (Pl. XLV, in pocket) outcrops can be found in stream beds that have been cut through the younger formations which elsewhere cover it. In some of the hills, especially those which are composed of basaltic lavas, such as the Bouse Hills and Palo Verde Hills, metamorphic rocks do not occur.

Many of the exposures of the basal complex consist of granitoid rocks. The bulk of these rocks are gray and pinkish gneisses which before their metamorphism were normal granites and intrusive rocks of similar types. These rocks are older than nearly all the other formations in the region, and they crop out in most of the mountain ranges. Plate XLI, *B*, shows their typical appearance. There are also in the region certain younger granites, not gneissoid, which Bancroft[1] considers to be Mesozoic. These are very similar in superficial appearance to the ancient granites in the several areas of such rocks mapped by previous workers. (See geologic map, Pl. XLV.) In the Buckskin Mountains near Osborne's Well there are outcrops of a fresh gray granite with no suggestion of gneissic structure. This rock contains specularite in places and perhaps has been otherwise mineralized, as several shallow prospect holes have been sunk in it. Probably it belongs to the group of Mesozoic intrusive rocks. The

boundaries of this mass were not mapped. Jones[2] reports the presence of Mesozoic intrusive rocks near Kofa, in the S. H. Mountains and in the Dome Rock Mountains. Bancroft[3] found dikes probably of Mesozoic age in the Harcuvar Mountains and Granite Wash Hills. It is probable that there are other areas of igneous rock of this age in the region.

The basal complex also includes small dikes composed for the most part of diabase and pegmatite. They are of general occurrence but have nowhere been found in large quantity. Bancroft[4] describes these rocks and also mentions certain exposures in the region north of that covered by the present report which he considers to be metamorphosed lava flows genetically related to the diabasic intrusive rocks.

In the Buckskin Mountains between Butler Well and Midway there are some good exposures of the ancient rocks. At one place in particular the unconformity between the major series of metamorphosed sediments, described below, and the gneiss can be clearly seen. Associated with the gneiss and clearly below the surface of unconformity are intensely metamorphosed schists, mostly somewhat chloritic. In the Gila Bend Mountains also there are small masses of fine-grained mica schists and quartzose schists included in the gneiss. At the southern extremity of the Big Horn Mountains, just north of the Palo Verde mine, is a hill composed entirely of dark-green foliated chloritic schist. (See Pl. XLI, *A*.) This rock is different from any observed elsewhere in the area, but presumably it is related in age to the rest of the metamorphic rocks. As it is very highly schistose, it is probably related to the most ancient of the schistose rocks. Blanchard[5] reports inclusions of metamorphosed limestone and dolomite in a few places in the gneiss of the Buckskin Mountains. In one outcrop of dolomite he found what he considers may be indistinct traces of organic remains.

[1] Bancroft, Howland, A reconnaissance of the ore deposits in northern Yuma County, Ariz.: U. S. Geol. Survey Bull. 451, p. 29, 1911.

[2] Jones, E. L., jr., A reconnaissance in the Kofa Mountains, Ariz.: U. S. Geol. Survey Bull. 620, pp. 151–164, 1916; Gold deposits near Quartzsite, Ariz.: Idem, p. 47.

[3] Bancroft, Howland, op. cit., p. 30.

[4] Idem, p. 28.

[5] Blanchard, R. C., The geology of the western Buckskin Mountains, Yuma County, Ariz.: Columbia Univ. Contr. Geol. Dept., vol. 26, No. 1, pp. 33–34, 1913.

A. HILLS OF CHLORITIC SCHIST AT THE NORTH END OF THE BIG HORN MOUNTAINS,
NEAR THE PALO VERDE MINE, MARICOPA COUNTY, ARIZ.

B. GONZALES WELLS, DOME ROCK MOUNTAINS, YUMA COUNTY, ARIZ.

The mountains in the background show the typical appearance of rocks of the basal complex.

A. A PLUG OF LATITE OF TERTIARY AGE IN THE DOME ROCK MOUNTAINS, ABOUT 4 MILES
SOUTHWEST OF QUARTZSITE, YUMA COUNTY, ARIZ.

B. BLACK BUTTE, CACTUS PLAIN, NEAR OSBORNE'S WELL, YUMA COUNTY, ARIZ.

An irregular intrusion of gabbro into Tertiary rocks.

The best exposures noted of the second type of metamorphic rocks of sedimentary origin are those in the Harcuvar, Harquahala, and Little Harquahala mountains, but they also occur in some of the other ranges. Interbedded limestones and quartzites are common, and the contrast in hardness between the rocks of these two types makes the bedding visible at considerable distances. The angles of dip in nearly all exposures noted are very moderate, and in most places the beds are nearly flat. Some of the lower beds in this series appear to be similar in composition to the underlying gneissic granite. They are doubtless metamorphosed arkosic sandstones derived from the ancient granitic rock. All the rocks of this series are notably metamorphosed, but many of them are not markedly schistose. Bancroft states that some of the calcareous rocks are dolomites. It is quite possible that more detailed work will result in further subdivision of this series. No attempt has been made to measure the thickness, but it is certainly considerably over 500 feet.

These rocks rest unconformably on the granitic gneisses which are so widespread in this region. The best exposure of the unconformity found is in the Buckskin Mountains on the road between Midway and Butler Well. At this locality gray granitic gneiss, inclosing masses of sericitic and chloritic schist, crops out. The schistosity strikes N. 30° W. and dips steeply to the southwest. These rocks are overlain by a mass of distinctly bedded but metamorphosed quartzite and sandstone, with some fine-grained crystalline limestone. The sandstone, especially near the base, is composed of débris from the gneiss below. Schistosity has been developed, especially in the sandstone, but it is not nearly so pronounced as it is in the lower rocks. The parting planes are parallel to the bedding. The beds have been crumpled, but not greatly. The maximum dip observed is 15°, and most of the beds are nearly flat. The average strike is N. 55° W., and the general dip is southwesterly. The section of the mass here exposed is over 150 feet thick. The contact between it and the gneiss and schist is somewhat irregular and is evidently an erosional contact. Both the older and the younger rocks are cut by small dikes of metamorphosed trap and by a dike composed entirely of microcline.

The known occurrences of the least metamorphosed sediments of the basal complex comprise those at the Socorro mine, in the Harquahala Mountains, and those in the northern portion of the Plomosa Mountains. Bancroft [6] has studied the section at the Socorro mine and gives the following description of it:

Coarse-grained granite which shows some schistosity is the basal rock in this locality and is similar to the pre-Cambrian granite so universally present in this area. Resting unconformably upon the granitic rock is a series of slightly metamorphosed sediments, of which about 150 feet of fine-grained grayish-red quartzite forms the base. This is overlain by several hundred feet of yellowish-brown limestone, the upper portion of which contains intercalated argillites and quartz-mica schists. Strata of schistose shaly limestone and a rock very closely resembling a dolomite (containing, however, fragments of quartz) were noticed near the contact of the quartzite and the overlying limestone.

In the Plomosa Mountains, near the Little Butte mine, there are limestones similar to those at the Socorro mine, but their relations to the underlying rocks were not determined. At no place were any of these comparatively slightly metamorphosed sedimentary rocks observed in contact with any of the highly metamorphosed sedimentary rocks in the region, and the relations between them were therefore not determined. The lithologic character of the least metamorphosed rocks is similar to that of some of the Paleozoic sedimentary rocks at Globe, Ariz. For this reason and because of their relation to rocks that are almost certainly pre-Cambrian and their comparatively small metamorphism, it seems probable that they are of Paleozoic age. The lack of fossils renders positive correlation impossible.

There can be little doubt that the granitic gneisses and associated metamorphosed sedimentary rocks just described, with the possible exception of the youngest of the sedimentary rocks, are of pre-Cambrian age. The fact that no fossils which can be used to determine the age of the beds have yet been found in any of the rocks examined during the present investigation makes all the determinations of the age of formation somewhat uncertain. However, it can not be questioned that these metamorphic rocks are very old. Some of them might conceivably be Paleozoic, but the

6 Bancroft, Howland, op. cit., pp. 111-112.

absence of fossils is a strong argument against this possibility, for most of the Paleozoic rocks of the region are fossiliferous. The fact that all these rocks except the youngest group are very much more metamorphosed than the known Paleozoic formations to the north and east is another strong reason for believing that they are pre-Cambrian rather than Paleozoic. There is no reason for believing that there has been more metamorphism in this area since Paleozoic time than has occurred in the Ray and Globe mining districts. The limestone and quartzite of the youngest group are not much if any more metamorphosed than similar rocks of Paleozoic age at Ray and Globe.

TERTIARY FORMATIONS.

GENERAL FEATURES.

Lavas occur throughout the area covered by this report and extend far beyond its limits. The series consists of a number of flows of varying thickness and of widely different superficial characteristics, associated with some tuffs and agglomerates and a very subordinate amount of sedimentary rock. It reaches its maximum development in the S. H. Mountains, where the total thickness is cerainly more than 2,000 feet. A number of the individual flows are several hundred feet thick.

Volcanic rocks similar in occurrence and general characteristics to rocks of this series have been reported from a number of localities in the Southwest. Such rocks are known in the Patagonia district, in southern Arizona;[7] in Mohave County, Ariz.,[8] to the north of the region covered by this report; in the Papago country,[9] just south of this region; in eastern California,[10] and in southern Nevada.[11] Similar rocks occur at Globe, in central Arizona,[12] and at many other places. These rocks have all been referred to the Tertiary, and most of

them are supposed to be Miocene. This supposition is based principally on their field relations to rocks of known age, the paleontologic evidence within the rocks themselves being scanty or lacking.

Overlying the Tertiary beds and associated with the unconsolidated or partly consolidated Quaternary sand and gravel are basalt flows of early Quaternary age. These will be discussed under the Quaternary formations. The faulted and uplifted basalts that cap many of the mountains, however, are considered to be of Tertiary age.

The amount of sedimentary material associated with the Tertiary lavas is small compared to the total thickness of the lavas. The sedimentary rocks are of geologic importance, however, for they furnish clues as to the conditions existing at the time these great flows occurred. They comprise sandstone, in part arkosic, shale, and calcareous beds.

TERTIARY LAVAS.

Distribution and character.—The Tertiary lavas are almost as universally present in this region as the metamorphic complex just described. They were found in almost every mountain range examined during this investigation, the only exceptions being the Harquahala and Little Harquahala mountains. Some of the ranges, such as the S. H., Eagle Tail, and Castle Dome mountains, are composed exclusively of rocks of this series resting on a metamorphic basement which is visible in only a few small areas.

The lavas are for the most part light-colored acidic rocks, but some are basalts. They display a wide range and variety of coloration. This is particularly striking in the Eagle Tail Mountains, where more than 1,000 feet of nearly horizontal lava flows, with interbedded tuff, is exposed. Nearly every flow is different in color from those above and below it, and each stands out from the others with clean-cut boundaries. Among the colors are brilliant yellow, soft green, vivid red, somber brown and dun, and creamy white, with streaks of purple, heliotrope, and other hues. The petrographer who is interested in Tertiary igneous rocks would find much to interest him here and in the other ranges in this region where similar rocks occur.

[7] Schrader, F. C., Mineral deposits of the Santa Rita and Patagonia mountains, Ariz.: U. S. Geol. Survey Bull. 582, pp. 70–76, 1915.

[8] Schrader, F. C., Mineral deposits of the Cerbat Range, Black Mountains, and Grand Wash Cliffs, Mohave County, Ariz.: U. S. Geol. Survey Bull. 340, pp. 57–59, 1907.

[9] Bryan, Kirk, The Papago country, Ariz.: U. S. Geol. Survey Water-Supply Paper — (in preparation).

[10] Brown, J. S., The Salton Sea region, Calif.: U. S. Geol. Survey Water-Supply Paper — (in preparation). Thompson, D. G., The Mohave Desert region, Calif.: U. S. Geol. Survey Water-Supply Paper — (in preparation).

[11] Ball, S. H., A geologic reconnaissance in southwestern Nevada and eastern California: U. S. Geol. Survey Bull. 308, pp. 31–34, 1907.

[12] Ransome, F. L., Geology of the Globe copper district, Ariz.: U. S. Geol. Survey Prof. Paper 12, pp. 88–95, 1903.

In most places the basalts appear to be the youngest of the flows, for they cap the others and form the summits of the mountains. Everywhere in the region the Tertiary basalts are subordinate in amount to the acidic flows. Thicknesses of 300 feet of basalt are rare, but 1,000 feet or more of acidic lava occurs at numerous places. The Tertiary basalts are best developed in the Gila Bend Mountains north of Point of Rocks.

Interbedded with the acidic flows are beds of siliceous agglomerate and rhyolitic tuff. The tuff is white or cream-colored, and the beds, which are in places scores of feet thick, are conspicuous. They have a wide distribution throughout the region.

The flows and tuffs are cut by pipes, dikes, and sills of felsitic igneous rock similar in general composition to the siliceous flows. The quantity of Tertiary intrusive rock exposed is very much less than that of the effusives. No large intrusive masses of this age are known anywhere in the region. A number of plugs or volcanic necks occur in the Plomosa, Dome Rock, and Eagle Tail mountains and some of the other ranges. A conspicuous plug in the Dome Rock Mountains is shown in Plate XLII, *A*. Court House Rock, a well-known landmark on the north side of the Eagle Tail Mountains, is a good example of such an intrusion. It is composed of cream-colored lava, in part weathered to a yellowish brown, and towers 1,000 feet sheer above its base, which is circular and only a few hundred feet in diameter. With the exception of a few cracks, mostly vertical, the walls are smooth and almost vertical nearly to the summit, where the cylindrical column has been partly broken by weathering. This peak is reported to have been scaled, truly a notable feat of mountain climbing. The range itself takes its name from a similar but even higher peak near its east end, whose summit is broken up into three points, showing a fancied resemblance to an eagle's tail sticking straight up into the air.

About 6 miles west of Osborne's Well, on the north side of one of the outlying hills of the Buckskin Mountains, is a scarp in which a peculiar exposure of igneous rock can be plainly seen. It is shown in Plate XLII, *B*. This is an intrusion of Tertiary age which differs in several respects from any seen elsewhere in the region. Microscopic examination shows

that the rock is a gabbro of coarse granulitic texture. The igneous mass has a very irregular outline, and the greatest extension exposed is in a horizontal direction. On the west are beds of brown sandstone dipping about 10° S. and striking roughly east. The contact with the gabbro is very irregular, and the sedimentary rocks are somewhat baked along it. Directly overlying the igneous rock is a basalt flow which caps the hill and is only 50 feet or so thick. When seen from a distance the lower part of the igneous mass seems to have a rough horizontal stratification, probably due to jointing. The upper part does not exhibit this apparent stratification but weathers in rounded masses 2 or 3 feet or more in diameter. The rock in these masses is full of grains of calcite, which give it a pseudoamygdaloidal appearance. The texture differs somewhat from that of the underlying portion, being on the whole coarser.

This irregular mass of gabbro was clearly intruded into the brown sandstone, which is almost certainly of Tertiary age. The basalt above is probably also Tertiary. There is no evidence to suggest that any other rock covered the basalt at the time the gabbro was intruded below it, but it is somewhat difficult to understand how a rock so coarsely crystalline as the gabbro could be intruded within 50 feet of the surface.

It should be noted that Bancroft[13] considered all the basalt in this part of Arizona to be Quaternary. Basalts occur on the summits of a number of mountains in the area. The amount of erosion since they were poured out is measured in thousands of feet, so that if these basalts are Pleistocene, some of the most imposing mountain ranges in the area have been produced in large part at least during later Pleistocene or Recent time. At Point of Rocks, on Gila River in the western part of Maricopa County, basalt flows capping unconsolidated gravel of the valley abut against the eroded edges of lava mountains. Hence, the basalt flows that cap these mountains must be older than the lava in the valley. As the latter caps unconsolidated gravel it is clearly Quaternary, and it is so greatly dissected by erosion and so much weathered that it is clearly early Pleistocene. From these facts it is evident that the older basalt capping the mountains belongs to the Tertiary. In many

13 Bancroft, Howland, op. cit., pp. 32–33.

places, however, it is very difficult or impossible to determine the age of a particular flow.

Petrography. — Petrographic examination shows that there is considerable similarity in type in these lavas throughout the area. Most of the flows and most of the intrusive rocks that cut them and are associated with them are latites and quartz latites; some are soda rhyolites. The tuffs examined are rhyolitic. Associated with these siliceous and sodic rocks, especially in the upper part of the series, are flows and dikes of basalt.

The latites are fine-grained rocks, in places porphyritic and commonly showing flow structure and perlitic growths. They are composed essentially of alkali plagioclase and orthoclase, with hornblende, biotite, and quartz usually present in subordinate amounts. Apatite and epidote were also noted in some specimens.

Most of the specimens of basalt examined are of the usual types, composed essentially of calcic plagioclase, augite, and olivine, with subordinate amounts of magnetite. They are somewhat porphyritic, all the minerals mentioned above, except the magnetite, occurring to a greater or less extent as phenocrysts. The groundmass is a fine-grained mass of plagioclase laths, showing in places parallel arrangement due to flowage, with granular augite and magnetite. Much of the olivine is altered to iddingsite.

Two specimens of basalt from the immediate vicinity of Woolsey Tank, in the Gila Bend Mountains, differ from those above described in that the feldspars are much more sodic. Their composition approaches that of oligoclase or albite-oligoclase. Stratigraphically these flows certainly are near the top of the series of Tertiary lavas, and they may even be of Pleistocene age. In the Buckskin Mountains near Osborne's Well Blanchard[14] found a rock which appears to be of a similar type.

TERTIARY SEDIMENTARY FORMATIONS.

Distribution and character. — Limestone and calcareous conglomerate occur in at least three widely separated localities in this area. Further work would probably disclose many other outcrops. The known localities are Osborne Wash, in the vicinity of Osborne's Well, near Parker; Saddle Mountain; and the Clanton Hills and the valley north of them. Sandstone was found in Antelope Hill, in several places in the Gila Bend Mountains, near Osborne's Well, in the Clanton Hills, and in small amounts elsewhere. Shale is associated with some of the sandstone in the Gila Bend Mountains.

Antelope Hill, at the south end of the concrete bridge across Gila River near Wellton, is composed of grayish arkose, a sandstone formed from granitic débris. The rock is, as a whole, somewhat coarser grained near the base of the hill than farther up the slope. The average diameter of the grains ranges from 1 to 6 millimeters. The beds have a very gentle southerly dip. The hill is about 580 feet high, so that fully 500 feet of sandstone is exposed. Related but coarser sandstone and conglomerate occur farther south.[15]

Red sandstone crops out in several places in the Gila Bend Mountains, notably at and near Woolsey Tank, where there is a bed 30 feet thick of sandstone interbedded with the limestone. Near the Dixie mine, in the Gila Bend Mountains, red and purplish shale is associated with the sandstone. Plate XLIV, *A* (p. 191), shows Tertiary sandstone in these mountains overlain by Pleistocene gravels.

The relations of these sedimentary rocks to the Tertiary lavas show clearly that they are of similar age. They have been disturbed, like the lavas, by post-Tertiary faulting, so that the beds now dip in various directions. The Clanton Hills, about 25 miles north of Palomas, consist almost exclusively of flat-lying gray cherty fine-grained limestone with numerous concretions, some of which resemble fossils in superficial appearance. Some of the beds contain small and indistinct fossils. (See pp. 189–190.) At the west end of the hills is exposed a bed of reddish sandstone composed of quartz grains in a calcareous cement, about 30 feet thick. There has been some faulting accompanied by considerable brecciation in the limestone. Subsequent to the faulting hot solutions circulated through the fault breccias, as is shown by iron stains and by marked silicification of the limestone fragments. No definite evidence of valuable mineralization was found.

---

[14] Blanchard, R. C., op. cit., pp. 26–27.

[15] Bryan, Kirk, The Papago country, Ariz.: U. S. Geol. Survey Water-Supply Paper — (in preparation).

A. OSBORNE WASH, ABOUT 2 MILES SOUTHWEST OF OSBORNE'S WELL, YUMA COUNTY, ARIZ.

A typical exposure of Tertiary limestone overlain by basalt.

B. SADDLE MOUNTAIN, MARICOPA COUNTY, ARIZ., LOOKING SOUTH.

Showing pockets or caves in Tertiary calcareous agglomerate or conglomerate.

A. BANK OF WASH NEAR WOOLSEY TANK, GILA BEND MOUNTAINS, MARICOPA COUNTY, ARIZ.

An exposure of tilted Tertiary sandstone unconformably overlain by tilted Quaternary gravel.

B. WOOLSEY TANK, GILA BEND MOUNTAINS, MARICOPA COUNTY, ARIZ.

Coarse gravel with calcareous cement, of Quaternary age, resting on the surface of Tertiary basalt.

Near Osborne's Well there are considerable exposures of sedimentary rocks. Time did not permit a detailed examination of them, but the scattered observations made may be of interest. To the west and south of the well are hills with cliffs cut by the large wash that passes between them. These cliffs expose conglomerate with a calcareous matrix, capped by a basalt flow. The calcareous rock is well bedded. The pebbles in it are in no place very abundant and in the lower portion are lacking altogether. Farther north up this wash are outcrops of red sandstone with concretions, a minor amount of quartz sandstone, and a few small beds of conglomerate. These rocks rest unconformably on gray granitic gneiss. The gneiss, which is very probably pre-Cambrian, is intruded by a light-colored granite, which looks very fresh and is not in the least gneissic in texture. This granite contains a little specularite and is apparently associated with certain small veins that contain quartz, specularite, and small amounts of sulphides and have been prospected to some extent in this vicinity. The granite, in common with similar rocks in the region associated with mineral veins, is probably Mesozoic. The sandstone rests unconformably on this granite, as well as on the older gneiss. A distinct though narrow bed of basal conglomerate containing granite pebbles occurs between the granite and the sandstone. A short distance farther north red vesicular basaltic or andesitic lava is interbedded with the red sandstone.

Exposures of sedimentary rocks are found for about 8 miles west of Osborne's Well along the road to Parker. There are numerous outcrops of thin-bedded limestone similar in appearance to the matrix of the conglomerate at the well, but entirely free from any but very small pebbles. Several of these outcrops are capped with vesicular olivine basalt. (See Pl. XLIII, A.) They contain rather numerous small and poorly preserved fossils (see next section) and a few small angular fragments of quartz and feldspar. Blanchard [16] considers these calcareous beds to be tuffaceous and states that the beds underlying the basalt at Osborne's Well have a groundmass of glass.

Interbedded with the lavas of Saddle Mountain, in Maricopa County, are considerable thicknesses of fragmental rocks ranging from agglomerate and breccia of distinctly igneous character to rocks that have angular fragments of lava about an inch in diameter in a white calcareous matrix. In certain cliffs there are peculiar hollows in beds of conglomerate and agglomerate, some of which almost amount to caves. (See Pl. XLIII, B.) The hollows appear to be due to a sort of concave exfoliation. They are not the result of solution or erosion.

Fossils.—The only fossils collected during this investigation were found in the limestone at two localities—in Osborne Wash near Osborne's Well and in the Clanton Hills. In Saddle Mountain there are beds of calcareous conglomerate which are lithologically very similar to those near Osborne's Well, but no specimens were collected from this locality, and it is therefore impossible to say whether these beds contain any small organic remains such as were found in the limestone near the well. The conglomerate of Saddle Mountain contains small bodies which are, superficially at least, similar to those in the limestone from the two localities mentioned, which W. H. Dall, of the United States Geological Survey, calls "pseudomorphs of what were probably a smooth cypridian crustacean."

Specimens from Osborne Wash and from the Clanton Hills were submitted to Mr. Dall for identification of the fossils. A specimen collected by John S. Brown, of the United States Geological Survey, from Imperial County, Calif., was submitted at the same time. This specimen, which closely resembles the specimen obtained near Osborne's Well, came from ledge in the west bank of an arroyo at the south entrance to a pass through the Palo Verde Mountains on the road from Glamis to Palo Verde, in either sec. 18 or 19, T. 10 S., R. 27 E., San Bernardino base line and meridian. Mr. Dall states that the specimen obtained near Osborne's Well and the one obtained by Mr. Brown in California "contain the same fossils and were doubtless laid down under practically identical conditions, whether absolutely contemporaneous or not." He found in these two specimens

small oval bodies representing pseudomorphs of what were probably a smooth cypridian crustacean, * * * also imprints of fragments of a gastropod which resemble analogous fragments of *Melania*, or *Goniobasis*, and a small triangular bivalve which appears to be most like a minute

---

[16] Blanchard, R. C., op. cit., pp. 24-26.

Corbicula, and not (as one might expect) belonging to the more common group of *Sphaerium* or *Pisidium*. There is also the imprint of a small leaf resembling a willow, and numerous lime tubes which seem to have been formed around roots or small vegetable stems.

Microscopic examination of the specimen found near Osborne's Well shows that it is very porous and is composed almost exclusively of calcareous matter in fragments of diverse shapes, with a few angular fragments of quartz and feldspar. The lime tubes mentioned by Mr. Dall are prominent in the thin section.

In a specimen of thin-bedded limestone from the east end of the Clanton Hills Mr. Dall found pseudomorphs of cypridian crustaceans like those he found in the other specimens mentioned. In one of the calcareous beds in Osborne Wash Blanchard [17] also found fossils, which Mr. Dall identified as the gastropod *Bittium* and a probable young *Corbicula*. He states: "There is nothing incompatible between the presence of *Bittium* with *Goniobasis* and *Corbicula* in the same deposit. All are prone to inhabit brackish water, especially near seashores." He also says: "There is no clue to the age of the deposit except that it is doubtless Tertiary."

QUATERNARY FORMATIONS.

SEDIMENTARY FORMATIONS.

Definition.—The unconsolidated and poorly consolidated gravel, sand, and silt that fill the valleys and floor the flood plains of the rivers in this desert region are of Quaternary age. Basalts which are clearly also Quaternary are interbedded with or rest upon these sediments.

Distribution and character.—The valleys throughout this area, like nearly all other desert valleys in the Southwest, are deeply filled with detrital material, for the most part unconsolidated or poorly consolidated, derived from the mountains. The thickness of this material in the valleys has not been determined. It is certainly to be measured in hundreds if not in thousands of feet, as is indicated by records of wells in a number of the valleys.

The character of the material varies greatly, as is to be expected in sediments laid down by generally short and usually disconnected streams under arid conditions. In the flood

plains of Gila and Colorado rivers and in certain clay flats, or playas, in interior valleys there are very fine silts or clays, but the major portion of the fill in the valleys is sand and gravel, in places very coarse. Much of it is poorly assorted, consisting of coarse sediments in a clayey matrix. The surface layers in most of the valleys contain silty soil more or less mixed with gravel. This soil, where it has been properly irrigated, has proved to be highly productive. In Castle Dome Plain, Palomas Plain, and to a less general extent in a number of the other valleys in the area the wind has removed the surface silt, leaving a residual floor of gravel. Sand dunes are common in Cactus Plain and also occur in Eagle Tail Valley.

In almost every place where the fill is indurated to any extent the cement is a calcareous material called "caliche" and known also as cement or hardpan. Lee [18] has described the mode of occurrence of caliche and discussed the theories as to its origin. He concludes that the caliche in Salt River valley, which is essentially similar to that in the lower Gila region, has been formed in part by the deposition of carbonates and other salts held in solution in the ground water and in part by the evaporation of water percolating downward from the surface. On the old road across the Gila Bend Mountains, west of Woolsey Tank, occur gravel beds with a calcareous cement which has set so firmly as to form a hard though friable rock. (See Pl. XLIV, *B*.) These beds are exceptionally indurated, but caliche beds so hard that it is very difficult to penetrate them with pick and shovel are common in a number of places in the region. Such beds are known elsewhere in the Gila Bend Mountains, in Nottbusch Valley, in Castle Dome Plain, and in other localities. Wells sunk in La Posa Plain and McMullen Valley usually penetrate beds of caliche below unconsolidated gravel and sand. On the flanks of the Plomosa Mountains, on the east side of La Posa Plain, lie thick deposits of caliche-cemented gravel, some of which is auriferous. [19] Similar deposits occur on the flanks of the Dome Rock Mountains west of this plain.

Beds of green and yellow banded clay are exposed in the terraces of Colorado River in the Colorado River Indian Reservation near Parker.

[17] Blanchard, R. C., op. cit., p. 39.

[18] Lee, W. T., Underground waters of Salt River valley, Ariz.: U. S. Geol. Survey Water-Supply Paper 136, pp. 107-111, 1905.

[19] Bancroft, Howland, op. cit., p. 88.

Fossil fresh-water shells have been found in some of these beds. E. L. Jones, jr.,[20] who made an examination of the reservation for the United States Geological Survey in 1914, states that these are lake beds.

Along washes within the mountains and on the borders of the ranges are beds of gravel and sand similar to those of the valley fill. These beds are cut by the present washes. Although they are clearly similar to the material in the modern streamways and were deposited under conditions very similar to those existing to-day, the position of many of these beds indicates that they were laid down in streams whose courses had little or no relation to those of the present streams. The gravel and sand are everywhere somewhat consolidated. In the wash that parallels the new road where it emerges from the mountains on the west side the unconsolidated or slightly consolidated gravel of the valley fill can be seen lapping up on the gently inclined and smooth surface of gravel with a calcareous cement. This gravel is continuous with gravel of the type just described occurring in the mountains proper. Similar exposures were noted near the road between Wenden and Butler Well, on the north side of Cunningham Pass, in the Harcuvar Mountains. Gravel of similar appearance, which is being eroded by the present streams, was noted in Osborne Wash, north of Osborne's Well, in the Buckskin Mountains.

The partly consolidated detrital beds in the mountains are in places cut by normal faults and tilted to angles of 20° and even more. The best exposures found are in the Gila Bend Mountains. (See Pl. XLIV, A.) Tilted blocks of gravel were noted near both of the roads that cross this range, but they are especially well exposed along the part of the old road that lies in the mountains. Outcrops of such material were found also along the large wash followed by the old road on the west side of the mountains. Slight folding in gravel beds was observed in some outcrops near Woolsey Tanks, along this road. Tilted gravel and sand are exposed at the north end of the Gila Mountains near Dome. Some of the more consolidated alluvium in the Dome Rock and Buckskin mountains is probably tilted. Beds of gravel and sand that have

been disturbed by earth movements doubtless exist elsewhere in the region but were not noted during this investigation.

It is evident that Quaternary sediments belonging to at least three periods of deposition occur in this area. These are (1) the somewhat consolidated beds exposed in and near the mountains, which have been disturbed by faulting, (2) the unconsolidated or only locally consolidated flat-lying valley fill, and (3) the recently deposited material in the washes and the playas of the desert valleys and in the flood plains of the larger streams. This conclusion is in accord with the results of Lee's work[21] in adjoining areas and in portions of the area here considered. He has given formational names to the two older Quaternary formations in the vicinity of Colorado River but not to the recent material flooring the river flood plains, etc. The oldest group of gravels and sands he calls the Temple Bar conglomerate. The unconsolidated material resting upon the Temple Bar conglomerate and exposed in terraced bluffs along Colorado River and elsewhere he calls the Chemehuevis gravel. The Temple Bar conglomerate is lithologically similar to the oldest of the three Quaternary formations herein described, but the thicknesses observed by Lee along the upper Colorado are far greater than any found in this region. The two may perhaps be of similar age and history. The Gila conglomerate, described by Gilbert,[22] is similar to the Temple Bar, being a thick formation of coarse alluvium in the upper Gila Valley. The correlation of these formations awaits the complete solution of the physiographic history of southwestern Arizona in Quaternary time.

QUATERNARY BASALT.

Associated with the gravel and sand of the valley fill in places in this area are flows of olivine basalt. Such rock caps the fill, is interbedded with it, and cuts it in the form of dikes and other intrusive masses, generally small and irregular. The basalt masses that rise above the present surface of the fill have produced

[20] Personal communication.

[21] Lee, W. T., Geologic reconnaissance of a part of western Arizona: U. S. Geol. Survey Bull. 352, pp. 17, 18, 1908; Underground waters of the Salt River valley, Ariz.: U. S. Geol. Survey Water-Supply Paper 136, pp. 111–114, 1905.

[22] Gilbert, G. K., U. S. Geog. and Geol. Surveys W. 100th Mer. Rept., vol. 3, pp. 540–541, 1875.

land forms of two general types. These are flat mesas formed by flows that have spread out over the surface of the fill, as at Point of Rocks and Enterprise dam, both on Gila River, and groups of low, more or less conical hills, of which the Bouse Hills, near Bouse, Yuma County, and the Palo Verde Hills, northwest of Arlington, Maricopa County, may be mentioned as examples. The mesas consist of flows 100 feet thick or less, with a few thicker ones. The hills are in general not over 200 or 300 feet high, and many are less than this. The conical shape of many of them suggests that they are volcanic plugs, but all are dissected by erosion, and nowhere was a definite crater found. All the basalt masses, in both mesas and hills, are eroded and have a weathered appearance. The basalt in this area is not nearly as fresh in appearance as much of that in California.[23] The relation of the basalt to the valley fill proves it to be Quaternary, but it is probably not younger than early Pleistocene.

STRUCTURE.

Normal faults are the most pronounced structural features of the rocks of this region. Thrust faults are not known anywhere in it, and only minor folds appear to have been formed since pre-Cambrian time. There seem to have been three general periods of faulting— (1) prior to the eruption of the Tertiary lava, (2) subsequent to the eruption of the Tertiary lava, and (3) subsequent to the deposition of the older Quaternary alluvium. These periods of movement are not sharply separated from one another. Indeed, it is probable that movement along fault planes has been in progress almost continuously from the beginning of pre-Tertiary faulting to the present day. A few of the mountain ranges in the region show no evidence of being faulted, either because they had a different origin or because erosion has entirely removed the evidence.

FOLDS.

The small blocks of early pre-Cambrian sedimentary rocks included in the gneiss in several localities are notably schistose. As regional schistosity can not be produced with-

[23] Darton, N. H., and others, Guidebook of the western United States, Part C, The Santa Fe Route: U. S. Geol. Survey Bull. 613, pp. 158 et seq., 1915.

out folding, such deformation must have taken place early in pre-Cambrian time. The later pre-Cambrian rocks are in large part not schistose, and over large areas their strata are flat or dip at gentle angles. Certainly no close folding has taken place in these strata since their deposition. The tipping of the beds in some localities is the result of faulting. As the rocks show evidence of widespread dynamic metamorphism they must have been subjected to great pressures, which probably resulted in broad and gentle doming.

In the Eagle Tail Mountains and probably in some of the other ranges the beds of Tertiary lava are curved in a way to suggest gentle local folds. This apparent bending may be and in most places probably is a result of original deposition and not of subsequent folding. Certainly no large amount of folding has affected the Tertiary lavas.

In the Gila Bend Mountains some of the beds of the older Quaternary alluvium have been gently folded, but most of the Quaternary deposits are undisturbed by folding.

FAULTS OLDER THAN THE TERTIARY LAVA.

When Tertiary volcanism began the surface of the region was irregular. Some of the mountain ranges which are present to-day existed then, although they may not have been as high or as rugged as they now are. The Harquahala and Harcuvar mountains and the Granite Wash Hills contain no known areas of lava and probably never were capped by such material. They are the result of some cause which antedates the lava. No evidence of close folding can be found in these ranges. It is possible, even probable, that their uplift was caused by faulting. There are several other ranges in the region that probably belong in this class, but so little is known about them that this is not certain.

The bold, almost precipitous northwestern face of the Harquahala Mountains has an appearance that suggests a fault scarp modified by erosion. The abrupt truncation of almost flat beds of pre-Cambrian sedimentary rocks in the southwestern slopes of the Granite Wash Hills near Vicksburg and elsewhere is also suggestive of faulting. Southwest of these hills, only a mile or so from their bases, are small hills of basalt of Tertiary or Quaternary age. The rock is more weathered than the basalts of known Quaternary age and is believed to be Tertiary. It was clearly erupted after the Granite Wash Hills came into existence.

Large portions of the Vulture Mountains, the White Tank Mountains in Maricopa County, the Palomas Mountains, the Maricopa Mountains, and other ranges lie above any known lava flows and probably never were covered by such flows. The character of the topography of areas of the basal complex from which the lavas have but recently been removed and the irregular lower contact of the lavas that are exposed in many places in the mountains show that the surface upon which they were poured out was hilly if not actually mountainous. This is especially well shown in the Gila Bend and Eagle Tail mountains.

In adjoining parts of Arizona there is conclusive evidence of pre-Tertiary faulting. Mountain building [24] attended by faulting took place during the Mesozoic era in what is now Cochise County. Bryan [25] has found evidence of a similar period of faulting in Pinal and Pima counties. Ransome [26] showed that faults of pre-Tertiary age probably exist in the Globe district, and this inference has been confirmed by later work in the Old Dominion mine.[27]

FAULTS YOUNGER THAN THE TERTIARY LAVA.

There is abundant evidence, both within and near the region covered by this report, of normal faulting subsequent to the eruption of the Tertiary lavas. The lavas have been broken into numerous blocks that dip in various directions and are bounded by faults with various strikes. Almost every range containing Tertiary lavas has obvious examples of such fault blocks. In some localities, as at Saddle Mountain, there are more or less heterogeneous groups of fault blocks. In others the whole mass of lavas composing the

range has been lifted, relative to the blocks on either side, with but little change in the horizontal attitude of the beds. The Eagle Tail Mountains and probably also the S. H. Mountains are of this character. The structure in both ranges is complicated by cross faults which have broken portions of the large block into smaller tilted blocks. Probably the Plomosa, Big Horn, Castle Dome, and other ranges are built up, in part or wholly, of such horst-like blocks broken by cross faults and carved by erosion, but the blocks of lava are now comparatively small, and many of them are tilted, so that the evidence as to the character of the originally dominant structure is obscure. Probably the faults of this period followed in part the lines of weakness developed during the pre-Tertiary crustal movements, but it is also probable that faulting along entirely new planes took place. Most of the ranges either trend approximately N. 50° W. or N. 50° E. or show a combination of these two directions. The strike of the ranges near Colorado River is more nearly north than that of most of those farther east. This is true both of those that strike west of north and those that strike east of it. The trends of the ranges doubtless correspond to the strikes of the major faults in them. Minor cross faulting in other directions also took place.

QUATERNARY FAULTS.

Probably no formations of known Quaternary age in this region are involved in large-scale faults. Minor earth movements broke and tilted the partly consolidated strata of older Quaternary alluvium in the Gila Bend Mountains and elsewhere. Probably some movement took place along the fault planes formed during the previous period of crustal disturbance. Lee [28] found evidence of considerable Quaternary faulting at Mesa and Tempe. This movement lowered some of the valley fill in this vicinity below sea level, as is shown by the log of a deep well at Mesa. Ransome [29] and others have shown that the Gila conglomerate is faulted in many places in the mountains east of Phoenix.

[24] Schrader, F. C., Mineral deposits of the Santa Rita and Patagonia Mountains, Ariz.: U. S. Geol. Survey Bull. 582, p. 77, 1915.

[25] Bryan, Kirk, The Papago country, Ariz.: U. S. Geol. Survey Water-Supply Paper — (in preparation).

[26] Ransome, F. L., Geology of the Globe copper district, Ariz.: U. S. Geol. Survey Prof. Paper 12, p. 104, 1903.

[27] Bjorge, G. N., personal communication.

[28] Lee, W. T., Underground waters of Salt River valley, Ariz.: U. S. Geol. Survey Water-Supply Paper 136, p. 115, 1905.

[29] Ransome, F. L., op. cit., p. 104.

GEOLOGIC HISTORY.

EARLY PRE-CAMBRIAN TIME.

The remnants of the oldest pre-Cambrian rocks in this area are so few, so widely scattered, and so intensely metamorphosed that almost nothing can be learned from them as to the events of that ancient time. The rocks referred to are the micaceous and chloritic schists, quartzitic schists, and metamorphosed limestones included in gneiss in the Buckskin and Gila Bend mountains. (See p. 184.) Some of these rocks have the megascopic appearance of highly altered sediments, but it is by no means certain that that is their origin. The large amount of chlorite in some of the schists suggests an igneous origin, but nothing more definite is known regarding them. All that the record shows is that in early pre-Cambrian time certain rocks, principally of sedimentary but perhaps also in part of igneous origin, existed here. These rocks were buried, metamorphosed, and finally intruded by batholithic masses of granite and kindred rocks. The period of intrusion was followed by a very long period of erosion. Nearly all the ancient schists were removed, and the granitic rock was exposed. Meanwhile the granites had been rendered gneissoid, and the blocks of other rocks included in them had been highly altered by dynamic metamorphism.

LATE PRE-CAMBRIAN TIME.

The next event recorded was sinking of the land and influx of the sea. A thick series of sandstone and limestone with some shale was laid down in this sea.

Many dikes, principally of diabase and pegmatite, are associated with the metamorphic formations. Some of these are to be correlated in age with the ancient batholithic intrusions and are older than the younger pre-Cambrian sedimentary rocks. Others clearly cut and are therefore younger than the later sedimentary rocks. The field work was not sufficiently detailed to make it possible to differentiate these dike rocks. Bancroft has found evidence in the northern portion of the area indicating that volcanism occurred during the period of marine sedimentation. (See p. 184.)

PALEOZOIC AND MESOZOIC TIME.

No sediments of known Paleozoic or Mesozoic age have been found in the region. Limestones and quartzites of possible Paleozoic age occur in the Harquahala Mountains and elsewhere. (See p. 185.) These rocks represent either sedimentation near the end of pre-Cambrian time or a continuation of marine sedimentation in the Paleozoic, but the evidence at hand is not sufficient to determine their age definitely. If any other Paleozoic or Mesozoic sediments were ever deposited in this region they have since been almost or entirely removed by erosion. It is possible that small amounts of such rocks occur in those parts of the region that were not visited during this investigation. Enough is known, however, to justify the belief that no large areas of such rocks are present anywhere in the region covered by this report.

The region was again uplifted at some time after the period of marine conditions recorded by the pre-Cambrian sedimentary rocks. Erosion was resumed and was long continued. If the marine sediments covered the whole region at the end of pre-Cambrian time, they have been completely removed over large areas and the gneisses once more laid bare. There is abundant evidence, however, that the surface over which the Tertiary lavas flowed was by no means a plain. The country was rolling and hilly. Some of the small mountain ranges that are present to-day existed then, although it is probable that they were not as high or as rugged as they are now.

Granitic stocks or small batholiths accompanied or immediately followed by dikes of various types were intruded into the rocks of this region at some period after the pre-Cambrian and before the Tertiary. The writers who have previously described such rocks consider them to be Mesozoic. This correlation seems to be probable and entirely in accord with the facts so far as they are known. Rocks of this type have been reported from the Dome Rock Mountains,[30] the S. H. Mountains,[31] and the Harcuvar Mountains[32] and

[30] Jones, E. L., jr., Gold deposits near Quartzsite, Ariz.: U. S. Geol. Survey Bull. 620, p. 48, 1916.
[31] Jones, E. L., jr., A reconnaissance in the Kofa Mountains, Ariz.: U. S. Geol. Survey Bull. 620, p. 155, 1916.
[32] Bancroft, Howland, Reconnaissance of the ore deposits in northern Yuma County, Ariz.: U. S. Geol. Survey Bull. 451, pp. 29–30, 1911.

were noted during the present investigation in the Buckskin Mountains. A number of similar intrusions are known in adjoining regions.

The pre-Cambrian rocks were considerably metamorphosed during the period between their deposition and that of the Tertiary lavas. The metamorphism probably took place in pre-Cambrian time, for Paleozoic rocks in adjoining regions show no evidence of having been affected by it. There has been no close folding since the deposition of the later pre-Cambrian rocks. Thick masses of these rocks are now exposed which show no folding and little tilting. Faulting took place at some period prior to the eruption of the Tertiary lavas, and it is probable that during that period the major areas of uplift which form the present areas of these rocks may have been blocked out, at least in part.

TERTIARY TIME.

The Tertiary was a period of pronounced volcanism, in which great sheets of lava were piled up in flow upon flow. Agglomerate and tuff are associated with the lavas, but in very subordinate amount. Quiet outflows rather than eruptions of explosive violence were the rule. Bancroft[33] states that volcanic plugs are present in several places in the area in northern Yuma County which he examined and are apparently more numerous near the lower part of Williams River than elsewhere. These plugs may represent remnants of Tertiary volcanoes. Plugs of latitic rock occur near Saddle Mountain, west of Quartzsite, in the Dome Rock Mountains, and at a few other places in the region covered by this report, but such remnants of Tertiary volcanoes are rare. Quite possibly most of the eruptions were of the fissure type, and no volcanoes, except a few small ones, ever existed here. Probably lava flowed over much of this region during the Tertiary period, covering most of the hills then existing. Apparently, however, some ranges were never capped completely by the lava. The Harquahala, Little Harquahala, and Harcuvar Mountains belong to this class, and portions of the Buckskin Mountains and of some of the other ranges may also have escaped being covered. Felsitic Tertiary intrusive rocks and possibly some lavas occur

in the Dome Rock Mountains, but this range consists almost exclusively of rocks of the basal complex. If the range was ever lava capped, all the lava has since been removed by erosion. Comparatively little is known in regard to the geology of the Laguna, Trigo, and Chocolate mountains. Possibly parts of these ranges escaped the general flooding of the region by the sheets of lava. Probably there was more than one period of extrusion, as has been found to be recorded elsewhere in similar rocks. Much more detailed work is required to determine this point.

The amount of sedimentary rocks of Tertiary age found in the area is small indeed compared to the many hundreds of feet of lavas. Unquestionably volcanism rather than sedimentation was the dominant feature of the period. Much of the Tertiary sedimentary rock is believed to be of terrestrial origin and was probably deposited under conditions not very different from those of the present day. This fact is better shown in the exposures of Tertiary formations south of the area covered by this guide, where stream-laid conglomerates occur.[34]

The calcareous sediments found in several places within this area and in adjoining parts of California tell a very different story. (See pp. 188–190.) These beds were unquestionably laid down in large bodies of quiet water. They are lacustrine or estuarine. A glance at the map will show that the exposures of these deposits are scattered over a region covering about 2,000 square miles. Only one of them, that near Osborne's Well, is in an area covered by an accurate topographic map, hence the exact altitudes of the others are not known. The best estimates available, however, show that all, including the California area, are at altitudes of approximately 700 feet above sea level. Unfortunately, the paleontologic evidence at hand is not conclusive as regards the character of the waters in which these beds were deposited. It is possible that they were laid down in lakes lying between the mountain ranges. Much more probably, however, they were deposited in an estuary or estuaries extending northward from the Gulf of California. In late Miocene or Pliocene time the gulf had a much greater extension to the north than at

33 Idem, pp. 30–31.

34 Bryan, Kirk, The Papago country, Ariz.: U. S. Geol. Survey Water-Supply Paper — (in preparation).

present, flooding southern California in the region of the Salton basin.[35] Possibly the calcareous beds in the region covered by this report mark the northern extension of this incursion of marine waters.

There was much normal faulting in Tertiary time, some of it on a large scale, and probably there was more than one period of faulting. It resulted in the formation of structural valleys between the upthrown blocks. Folding either did not occur or was of very minor amount.

QUATERNARY TIME.

Our knowledge of Quaternary events in this region is more detailed and complete than that of the older geologic periods. Doubtless there were several divisions of Tertiary time besides those mentioned above. The great masses of lava, for example, probably were not all poured out during one continuous period of eruption. There were interruptions and alternations of conditions. The record of these events is so fragmentary and obscure, however, that it was impossible to work out the details of the Tertiary history. The record of Quaternary events is naturally much more completely preserved, though there is much that is still uncertain or entirely unknown regarding the history of this period. One of the greatest difficulties encountered in interpreting the record is that of differentiating between the older and the younger valley fill, which are lithologically very similar.

Some uncertainty exists as to the division between the Tertiary and Quaternary in this region. Lee[36] believes that the uplift that initiated the cutting of the Grand Canyon of the Colorado marks the beginning of the Quaternary period. This uplift was very probably essentially contemporaneous with that which resulted in the deep cutting of the desert valleys. However, Lee elsewhere[37] makes the suggestion that the lower portion of the fill in Salt River valley may be Tertiary. He considers that this lower portion may be lacustrine in origin and notably

older than the detrital material above it. Deep wells show that there is a considerable thickness of clay or other fine material beneath the coarser detritus in Salt River valley. Records of wells in Buckeye and Arlington valleys and at Gila Bend show that similar conditions exist in those localities also. Considerable clay was encountered in several of the Southern Pacific Railroad wells on the Gila west of Gila Bend. It is possible that fossil or other evidence may eventually be found which will show that these beds are Tertiary, but until further facts are discovered the most logical conclusion appears to be to consider the deep cutting of the valleys, originally in large part of structural origin, as the first event of the Quaternary period in the region under discussion. Any sediments, whatever their origin, lying in these valleys must then be considered of Quaternary age. The mere fact that the lower part of the fill is apparently of lacustrine origin does not affect the problem of its age. Beds of unquestionably Quaternary age and very probably lacustrine origin occur near Parker, on Colorado River. A temporary lake[38] is believed to have existed in Arlington Valley in recent geologic time.

After the valley cutting the conditions were so altered that the streams began to aggrade, and the recently excavated valleys were filled to great depths with detrital material. Basalt flows, the continuation of the basaltic effusions at the end of the Tertiary, occurred at this time. As has already been stated, volcanism did not continue as late in this region as it did in some other portions of the Southwest, notably southern California. It continued intermittently to a time considerably later than that in which the first valley fill was deposited.

When the valleys had been very largely filled with detritus, renewed uplift occurred. In places the recently deposited sediments were faulted and somewhat folded. Degradation recommenced, and much of the material with which the valleys had just been filled was swept out of them.

Before all of the first valley fill had been removed aggradation was resumed and the

[35] Kew, W. S. W., Tertiary echinoids of the Carrizo Creek region in the Colorado Desert: California Univ. Dept. Geology Bull., vol. 8, pp. 39-60, 1914.

[36] Lee, W. T., Geologic reconnaissance of a part of western Arizona: U. S. Geol. Survey Bull. 352, pp. 62-63, 1908.

[37] Lee, W. T., Underground waters of Salt River valley, Ariz., U. S. Geol. Survey Water-Supply Paper 136, p. 114, 1905.

[38] Ross, C. P., The lower Gila region, Ariz.: U. S. Geol. Survey Water-Supply Paper — (in preparation).

younger fill was deposited. Volcanism of minor extent occurred during this epoch. Near Bouse, Yuma County, volcanic ash occurs in the fill not far from the present surface. This deposit is probably comparatively recent. Several of the lava flows may be of corresponding age.

In comparatively recent time erosion of the younger fill has commenced, as is shown by terraces cut in it. The present flood plains of the streams lie between the lowest of these terraces. Along both Colorado and Gila rivers other terraces can be discerned above these, but they are discontinuous and apparently not significant.

At the present time both rivers are aggrading in their lower courses. Their channels are gradually being filled by the deposition of fine silts. Both rivers are well known for the large quantities of silt carried by their waters during floods. At all times they are remarkably muddy.

MINERAL DEPOSITS.

This part of Arizona has been extensively prospected. Mineral deposits are now known to occur in every mountain range and in many of the groups of hills within the region. The only hills in which mineral deposits have not been, and in all probability will not be found, are those composed exclusively of Quaternary basalt.

The types of deposits and the minerals found are many and diverse. Mining has been carried on in this region for gold, silver, copper, lead, zinc, mercury, iron, and manganese. There has been some prospecting for tungsten, but no mining for this metal has been done. Fluorite occurs in the Castle Dome district and possibly elsewhere but so far as known has not been developed. Gypsum occurs in some places in the region, but no deposits of commercial importance are known.

Mining is in progress in several of the mountain ranges in this region, but there are no large mines operating at present. In the past the Vulture mine, in the range of the same name; the mines about Kofa, in the S. H. Mountains; the Harqua Hala or Bonanza mine, in the Little Harquahala Mountains; and other less well-known properties have shipped considerable gold ore. Silver and lead were mined for some years in the Castle Dome Mountains.

Gold placers were formerly worked along Colorado River near La Paz and Ehrenberg, in and near the Dome Rock and Plomosa Mountains, and at Gila City on Gila River, at the site of the present town of Dome. Some placer mining is still in progress in the Plomosa and Dome Rock mountains, but elsewhere activity of this sort has almost entirely ceased. The lack of water appears to be the principal obstacle to the successful development of the placers. In the old vein mines the richer and more accessible portions of the ore bodies have been worked out, and lack of transportation facilities and of capital have prevented further development. The Harqua Hala mine is now being reopened for copper. It is possible that many of the mines now abandoned could again be made profitable producers by extending the workings deeper.

At present (1918) there is considerable activity in the small copper mines in the vicinity of Cunningham Pass, in the Harcuvar Mountains. Mining for copper and other metals is being carried on in the Buckskin and Plomosa mountains and to a small extent elsewhere. More or less desultory prospecting is in progress in all the mountain ranges. In 1918 plans were being discussed for a reopening of some of the mines in the vicinity of Kofa.

A detailed discussion of the geologic features of the ore deposits in the region is beyond the scope of this paper, but descriptions of some of the mines will be found in the final report [39] now being prepared. Bancroft [40] has described the deposits in the northern part of the region, and the deposits noted in the southern part appear to be in general similar to the types he describes. According to him there were three periods of mineralization—pre-Cambrian, Mesozoic, and Tertiary. He describes numerous types of deposits belonging to these periods, also the placers in the vicinity of Quartzsite. The reports of the governor of the Territory of Arizona contain many references by W. P. Blake, Territorial geologist, to deposits in this area. Jones has described deposits in the Dome Rock Mountains [41] and near Kofa, in the S. H. Mountains.[42]

[39] Ross, C. P., The lower Gila region, Ariz.: U. S. Geol. Survey Water-Supply Paper — (in preparation).
[40] Bancroft, Howland, op. cit.
[41] Jones, E. L., jr., Gold deposits near Quartzsite, Ariz.: U. S. Geol. Survey Bull. 620, pp. 45-57, 1916.
[42] Jones, E. L., jr., A reconnaissance in the Kofa Mountains, Ariz.: U. S. Geol. Survey Bull. 620, pp. 151-164, 1916.

GOLD RUSH BOOKS

OREGON, USA

www.GoldMiningBooks.com

Books On Mining

Visit: www.goldminingbooks.com to order your copies or ask your favorite book seller to offer them.

Mining Books by Kerby Jackson

<u>Gold Dust: Stories From Oregon's Mining Years</u> - Oregon mining historian and prospector, Kerby Jackson, brings you a treasure trove of seventeen stories on Southern Oregon's rich history of gold prospecting, the prospectors and their discoveries, and the breathtaking areas they settled in and made homes. **5" X 8", 98 ppgs. Retail Price: $11.99**

<u>The Golden Trail: More Stories From Oregon's Mining Years</u> - In his follow-up to "Gold Dust: Stories of Oregon's Mining Years", this time around, Jackson brings us twelve tales from Oregon's Gold Rush, including the story about the first gold strike on Canyon Creek in Grant County, about the old timers who found gold by the pail full at the Victor Mine near Galice, how Iradel Bray discovered a rich ledge of gold on the Coquille River during the height of the Rogue River War, a tale of two elderly miners on the hunt for a lost mine in the Cascade Mountains, details about the discovery of the famous Armstrong Nugget and others. **5" X 8", 70 ppgs. Retail Price: $10.99**

Oregon Mining Books

<u>Geology and Mineral Resources of Josephine County, Oregon</u> - Unavailable since the 1970's, this important publication was originally compiled by the Oregon Department of Geology and Mineral Industries and includes important details on the economic geology and mineral resources of this important mining area in South Western Oregon. Included are notes on the history, geology and development of important mines, as well as insights into the mining of gold, copper, nickel, limestone, chromium and other minerals found in large quantities in Josephine County, Oregon. **8.5" X 11", 54 ppgs. Retail Price: $9.99**

<u>Mines and Prospects of the Mount Reuben Mining District</u> - Unavailable since 1947, this important publication was originally compiled by geologist Elton Youngberg of the Oregon Department of Geology and Mineral Industries and includes detailed descriptions, histories and the geology of the Mount Reuben Mining District in Josephine County, Oregon. Included are notes on the history, geology, development and assay statistics, as well as underground maps of all the major mines and prospects in the vicinity of this much neglected mining district. **8.5" X 11", 48 ppgs. Retail Price: $9.99**

<u>The Granite Mining District</u> - Notes on the history, geology and development of important mines in the well known Granite Mining District which is located in Grant County, Oregon. Some of the mines discussed include the Ajax, Blue Ribbon, Buffalo, Continental, Cougar-Independence, Magnolia, New York, Standard and the Tillicum. Also included are many rare maps pertaining to the mines in the area. **8.5" X 11", 48 ppgs. Retail Price: $9.99**

<u>Ore Deposits of the Takilma and Waldo Mining Districts of Josephine County, Oregon</u> - The Waldo and Takilma mining districts are most notable for the fact that the earliest large scale mining of placer gold and copper in Oregon took place in these two areas. Included are details about some of the earliest large gold mines in the state such as the Llano de Oro, High Gravel, Cameron, Platerica, Deep Gravel and others, as well as copper mines such as the famous Queen of Bronze mine, the Waldo, Lily and Cowboy mines. This volume also includes six maps and 20 original illustrations. **8.5" X 11", 74 ppgs. Retail Price: $9.99**

<u>Metal Mines of Douglas, Coos and Curry Counties, Oregon</u> - Oregon mining historian Kerby Jackson introduces us to a classic work on Oregon's mining history in this important re-issue of Bulletin 14C Volume 1, otherwise known as the Douglas, Coos & Curry Counties, Oregon Metal Mines Handbook. Unavailable since 1940, this important publication was originally compiled by the Oregon Department of Geology and Mineral Industries includes detailed descriptions, histories and the geology of over 250 metallic mineral mines and prospects in this rugged area of South West Oregon. **8.5" X 11", 158 ppgs. Retail Price: $19.99**

Metal Mines of Jackson County, Oregon - Unavailable since 1943, this important publication was originally compiled by the Oregon Department of Geology and Mineral Industries includes detailed descriptions, histories and the geology of over 450 metallic mineral mines and prospects in Jackson County, Oregon. Included are such famous gold mining areas as Gold Hill, Jacksonville, Sterling and the Upper Applegate. **8.5" X 11", 220 ppgs. Retail Price: $24.99**

Metal Mines of Josephine County, Oregon - Oregon mining historian Kerby Jackson introduces us to a classic work on Oregon's mining history in this important re-issue of Bulletin 14C, otherwise known as the Josephine County, Oregon Metal Mines Handbook. Unavailable since 1952, this important publication was originally compiled by the Oregon Department of Geology and Mineral Industries includes detailed descriptions, histories and the geology of over 500 metallic mineral mines and prospects in Josephine County, Oregon. **8.5" X 11", 250 ppgs. Retail Price: $24.99**

Metal Mines of North East Oregon - Oregon mining historian Kerby Jackson introduces us to a classic work on Oregon's mining history in this important re-issue of Bulletin 14A and 14B, otherwise known as the North East Oregon Metal Mines Handbook. Unavailable since 1941, this important publication was originally compiled by the Oregon Department of Geology and Mineral Industries and includes detailed descriptions, histories and the geology of over 750 metallic mineral mines and prospects in North Eastern Oregon. **8.5" X 11", 310 ppgs. Retail Price: $29.99**

Metal Mines of North West Oregon - Oregon mining historian Kerby Jackson introduces us to a classic work on Oregon's mining history in this important re-issue of Bulletin 14D, otherwise known as the North West Oregon Metal Mines Handbook. Unavailable since 1951, this important publication was originally compiled by the Oregon Department of Geology and Mineral Industries and includes detailed descriptions, histories and the geology of over 250 metallic mineral mines and prospects in North Western Oregon. **8.5" X 11", 182 ppgs. Retail Price: $19.99**

Mines and Prospects of Oregon - Mining historian Kerby Jackson introduces us to a classic mining work by the Oregon Bureau of Mines in this important re-issue of The Handbook of Mines and Prospects of Oregon. Unavailable since 1916, this publication includes important insights into hundreds of gold, silver, copper, coal, limestone and other mines that operated in the State of Oregon around the turn of the 19th Century. Included are not only geological details on early mines throughout Oregon, but also insights into their history, production, locations and in some cases, also included are rare maps of their underground workings. **8.5" X 11", 314 ppgs. Retail Price: $24.99**

Lode Gold of the Klamath Mountains of Northern California and South West Oregon
(See California Mining Books)

Mineral Resources of South West Oregon - Unavailable since 1914, this publication includes important insights into dozens of mines that once operated in South West Oregon, including the famous gold fields of Josephine and Jackson Counties, as well as the Coal Mines of Coos County. Included are not only geological details on early mines throughout South West Oregon, but also insights into their history, production and locations. **8.5" X 11", 154 ppgs. Retail Price: $11.99**

Chromite Mining in The Klamath Mountains of California and Oregon
(See California Mining Books)

Southern Oregon Mineral Wealth - Unavailable since 1904, this rare publication provides a unique snapshot into the mines that were operating in the area at the time. Included are not only geological details on early mines throughout South West Oregon, but also insights into their history, production and locations. Some of the mining areas include Grave Creek, Greenback, Wolf Creek, Jump Off Joe Creek, Granite Hill, Galice, Mount Reuben, Gold Hill, Galls Creek, Kane Creek, Sardine Creek, Birdseye Creek, Evans Creek, Foots Creek, Jacksonville, Ashland, the Applegate River, Waldo, Kerby and the Illinois River, Althouse and Sucker Creek, as well as insights into local copper mining and other topics. **8.5" X 11", 64 ppgs. Retail Price: $8.99**

Geology and Ore Deposits of the Takilma and Waldo Mining Districts - Unavailable since the 1933, this publication was originally compiled by the United States Geological Survey and includes details on gold and copper mining in the Takilma and Waldo Districts of Josephine County, Oregon. The Waldo and Takilma mining districts are most notable for the fact that the earliest large scale mining of placer gold and copper in Oregon took place in these two areas. Included in this report are details about some of the earliest large gold mines in the state such as the Llano de Oro, High Gravel, Cameron, Platerica, Deep Gravel and others, as well as copper mines such as the famous Queen of Bronze mine, the Waldo, Lily and Cowboy mines. In addition to geological examinations, insights are also provided into the production, day to day operations and early histories of these mines, as well as calculations of known mineral reserves in the area. This volume also includes six maps and 20 original illustrations. **8.5" X 11", 74 ppgs. Retail Price: $9.99**

Gold Mines of Oregon - Oregon mining historian Kerby Jackson introduces us to a classic work on Oregon's mining history in this important re-issue of Bulletin 61, otherwise known as "Gold and Silver In Oregon". Unavailable since 1968, this important publication was originally compiled by geologists Howard C. Brooks and Len Ramp of the Oregon Department of Geology and Mineral Industries and includes detailed descriptions, histories and the geology of over 450 gold mines Oregon. Included are notes on the history, geology and gold production statistics of all the major mining areas in Oregon including the Klamath Mountains, the Blue Mountains and the North Cascades. While gold is where you find it, as every miner knows, the path to success is to prospect for gold where it was previously found. 8.5" X 11", 344 ppgs. Retail Price: $24.99

Mines and Mineral Resources of Curry County Oregon - Originally published in 1916, this important publication on Oregon Mining has not been available for nearly a century. Included are rare insights into the history, production and locations of dozens of gold mines in Curry County, Oregon, as well as detailed information on important Oregon mining districts in that area such as those at Agness, Bald Face Creek, Mule Creek, Boulder Creek, China Diggings, Collier Creek, Elk River, Gold Beach, Rock Creek, Sixes River and elsewhere. Particular attention is especially paid to the famous beach gold deposits of this portion of the Oregon Coast. 8.5" X 11", 140 ppgs. Retail Price: $11.99

Chromite Mining in South West Oregon - Originally published in 1961, this important publication on Oregon Mining has not been available for nearly a century. Included are rare insights into the history, production and locations of nearly 300 chromite mines in South Western Oregon. 8.5" X 11", 184 ppgs. Retail Price: $14.99

Mineral Resources of Douglas County Oregon - Originally published in 1972, this important publication on Oregon Mining has not been available for nearly forty years. Included are rare insights into the geology, history, production and locations of numerous gold mines and other mining properties in Douglas County, Oregon. 8.5" X 11", 124 ppgs. Retail Price: $11.99

Mineral Resources of Coos County Oregon - Originally published in 1972, this important publication on Oregon Mining has not been available for nearly forty years. Included are rare insights into the geology, history, production and locations of numerous gold mines and other mining properties in Coos County, Oregon. 8.5" X 11", 100 ppgs. Retail Price: $11.99

Mineral Resources of Lane County Oregon - Originally published in 1938, this important publication on Oregon Mining has not been available for nearly seventy five years. Included are extremely rare insights into the geology and mines of Lane County, Oregon, in particular in the Bohemia, Blue River, Oakridge, Black Butte and Winberry Mining Districts. 8.5" X 11", 82 ppgs. Retail Price: $9.99

Mineral Resources of the Upper Chetco River of Oregon: Including the Kalmiopsis Wilderness - Originally published in 1975, this important publication on Oregon Mining has not been available for nearly forty years. Withdrawn under the 1872 Mining Act since 1984, real insight into the minerals resources and mines of the Upper Chetco River has long been unavailable due to the remoteness of the area. Despite this, the decades of battle between property owners and environmental extremists over the last private mining inholding in the area has continued to pique the interest of those interested in mining and other forms of natural resource use. Gold mining began in the area in the 1850's and has a rich history in this geographic area, even if the facts surrounding it are little known. Included are twenty two rare photographs, as well as insights into the Becca and Morning Mine, the Emmly Mine (also known as Emily Camp), the Frazier Mine, the Golden Dream or Higgins Mine, Hustis Mine, Peck Mine and others. 8.5" X 11", 64 ppgs. Retail Price: $8.99

Gold Dredging in Oregon - Originally published in 1939, this important publication on Oregon Mining has not been available for nearly seventy five years. Included are extremely rare insights into the history and day to day operations of the dragline and bucketline gold dredges that once worked the placer gold fields of South West and North East Oregon in decades gone by. Also included are details into the areas that were worked by gold dredges in Josephine, Jackson, Baker and Grant counties, as well as the economic factors that impacted this mining method. This volume also offers a unique look into the values of river bottom land in relation to both farming and mining, in how farm lands were mined, re-soiled and reclamated after the dredges worked them. Featured are hard to find maps of the gold dredge fields, as well as rare photographs from a bygone era. 8.5" X 11", 86 ppgs. Retail Price: $8.99

Quick Silver Mining in Oregon - Originally published in 1963, this important publication on Oregon Mining has not been available for over fifty years. This publication includes details into the history and production of Elemental Mercury or Quicksilver in the State of Oregon. 8.5" X 11", 238 ppgs. Retail Price: $15.99

Mines of the Greenhorn Mining District of Grant County Oregon - Originally published in 1948, this important publication on Oregon Mining has not been available for over sixty five years. In this publication are rare insights into the mines of the famous Greenhorn Mining District of Grant County, Oregon, especially the famous Morning Mine. Also included are details on the Tempest, Tiger, Bi-Metallic, Windsor, Psyche, Big Johnny, Snow Creek, Banzette and Paramount Mines, as well as prospects in the vicinities in the famous mining areas of Mormon Basin, Vinegar Basin and Desolation Creek. Included are hard to find mine maps and dozens of rare photographs from the bygone era of Grant County's rich mining history. 8.5" X 11", 72 ppgs. Retail Price: $9.99

Geology of the Wallowa Mountains of Oregon: Part I (Volume 1) - Originally published in 1938, this important publication on Oregon Mining has not been available for nearly seventy five years. Included are details on the geology of this unique portion of North Eastern Oregon. This is the first part of a two book series on the area. Accompanying the text are rare photographs and historic maps.**8.5" X 11", 92 ppgs. Retail Price: $9.99**

Geology of the Wallowa Mountains of Oregon: Part II (Volume 2) - Originally published in 1938, this important publication on Oregon Mining has not been available for nearly seventy five years. Included are details on the geology of this unique portion of North Eastern Oregon. This is the first part of a two book series on the area. Accompanying the text are rare photographs and historic maps.**8.5" X 11", 94 ppgs. Retail Price: $9.99**

Field Identification of Minerals For Oregon Prospectors - Originally published in 1940, this important publication on Oregon Mining has not been available for nearly seventy five years. Included in this volume is an easy system for testing and identifying a wide range of minerals that might be found by prospectors, geologists and rockhounds in the State of Oregon, as well as in other locales. Topics include how to put together your own field testing kit and how to conduct rudimentary tests in the field. This volume is written in a clear and concise way to make it useful even for beginners. **8.5" X 11", 158 ppgs. Retail Price: $14.99**

The Bohemia Mining District of Oregon - Originally published in 1900, this important publication on Oregon Mining has not been available for over a century. Included in this volume are important insights into the famous Bohemia Mining District of Oregon, including the histories and locations of important gold mines in the area such as the Ophir Mine, Clarence, Acturas, Peek-a-boo, White Swan, Combination Mine, the Musick Mine, The California, White Ghost, The Mystery, Wall Street, Vesuvius, Story, Lizzie Bullock, Delta, Elsie Dora, Golden Slipper, Broadway, Champion Mine, Knott, Noonday, Helena, White Wings, Riverside and others. Also included are notes on the nearby Blue River Mining District. **8.5" X 11", 58 ppgs. Retail Price: $9.99**

The Gold Fields of Eastern Oregon - Unavailable since 1900, this publication was originally compiled by the Baker City Chamber of Commerce Offering important insights into the gold mining history of Eastern Oregon, "The Gold Fields of Eastern Oregon" sheds a rare light on many of the gold mines that were operating at the turn of the 19th Century in Baker County and Grant County in North Eastern Oregon. Some of the areas featured include the Cable Cove District, Baisely-Elhorn, Granite, Red Boy, Bonanza, Susanville, Sparta, Virtue, Vaughn, Sumpter, Burnt River, Rye Valley and other mining districts. Included is basic information on not only many gold mines that are well known to those interested in Eastern Oregon mining history, but also many mines and prospects which have been mostly lost to the passage of time. Accompanying are numerous rare photos **8.5" X 11", 78 ppgs. Retail Price: $10.99**

Gold Mining in Eastern Oregon - Originally published in 1938, this important publication on Oregon Mining has not been available for over a century. Included in this volume are important insights into the famous mining districts of Eastern Oregon during the late 1930's. Particular attention is given to those gold mines with milling and concentrating facilities in the Greenhorn, Red Boy, Alamo, Bonanza, Granite, Cable Cove, Cracker Creek, Virtue, Keating, Medical Springs, Sanger, Sparta, Chicken Creek, Mormon Basin, Connor Creek, Cornucopia and the Bull Run Mining Districts. Some of the mines featured include the Ben Harrison, North Pole-Columbia, Highland Maxwell, Baisley-Elkhorn, White Swan, Balm Creek, Twin Baby, Gem of Sparta, New Deal, Gleason, Gifford-Johnson, Cornucopia, Record, Bull Run, Orion and others. Of particular interest are the mill flow sheets and descriptions of milling operations of these mines. **8.5" X 11", 68 ppgs. Retail Price: $8.99**

The Gold Belt of the Blue Mountains of Oregon - Originally published in 1901, this important publication on Oregon Mining has not been available for over a century. Included in this volume are rare insights into the gold deposits of the Blue Mountains of North East Oregon, including the history of their early discovery and early production. Extensive details are offered on this important mining area's mineralogy and economic geology, as well as insights into nearby gold placers, silver deposits and copper deposits. Featured are the Elkhorn and Rock Creek mining districts, the Pocahontas district, Auburn and Minersville districts, Sumpter and Cracker Creek, Cable Cove, the Camp Carson district, Granite, Alamo, Greenhorn, Robinsonville, the Upper Burnt River Valley and Bonanza districts, Susanville, Quartzburg, Canyon Creek, Virtue, the Copper Butte district, the North Powder River, Sparta, Eagle Creek, Cornucopia, Pine Creek, Lower Powder River, the Upper Snake River Canyon, Rye Valley, Lower Burnt River Valley, Mormon Basin, the Malheur and Clarks Creek districts, Sutton Creek and others. Of particular interest are important details on numerous gold mines and prospects in these mining districts, including their locations, histories, geology and other important information, as well as information on silver, copper and fire opal deposits. **8.5" X 11", 250 ppgs. Retail Price: $24.99**

Mining in the Cascades Range of Oregon - Originally published in 1938, this important publication on Oregon Mining has not been available for over seventy five years. Included in this volume are rare insights into the gold mines and other types of metal mines in the Cascades Mountain Range of Oregon. Some of the important mining areas covered include the famous Bohemia Mining District, the North Santiam Mining District, Quartzville Mining District, Blue River Mining District, Fall Creek Mining District, Oakridge District, Zinc District, Buzzard-Al Sarena District, Grand Cove, Climax District and Barron Mining District. Of particular interest are important details on over 100 mines and prospects in these mining districts, including their locations, histories, geology and other important information. **8.5" X 11", 170 ppgs. Retail Price: $14.99**

Beach Gold Placers of the Oregon Coast - Originally published in 1934, this important publication on Oregon Mining has not been available for over 80 years. Included in this volume are rare insights into the beach gold deposits of the State of Oregon, including their locations, occurance, composition and geology. Of particular interest is information on placer platinum in Oregon's rich beach deposits. Also included are the locations and other information on some famous Oregon beach mines, including the Pioneer, Eagle, Chickamin, Iowa and beach placer mines north of the mouth of the Rogue River. 8.5" X 11", 60 ppgs. Retail Price: $8.99

Mineralogical Composition of the Sands of the Oregon Coast: From Coos Bay to the Columbia - Published in 1945, he text features hard to find information on the composition of the gold bearing black sands of the South West Oregon Coast, offering a unique insight to prospectors in search of Oregon's legendary beach gold. 104 ppgs, $9.99

Manganese Mining in Oregon - First released in 1942 and now out of print, this special reprint edition of "Manganese in Oregon" was originally published by the Oregon Department of Geology and Mineral Industries. The text features hard to find information on the mining of Manganese in Oregon, including details and maps of Oregon manganese mines and prospects. 108 ppgs, 9.99

Medford Oregon As A Mining Center - Written in 1912, this hard to find publication includes valuable insights into the mining history of South West Oregon. This small book contains interesting information on the gold, copper and mining industry in Southern Oregon as it existed just prior to World War One, shedding light on some of the important mines in the area. Included are rare photographs and vintage advertising of the day. 80 ppgs, 9.99

Mineral Resources of Curry County Oregon - First released in 1977 and now out of print, this special reprint edition of "Geology, Mineral Resources and Rock Materials of Curry County, Oregon" was originally published in cooperation of Curry County, Oregon and the Oregon Department of Geology and Mineral Industries. The text features hard to find information on not only the mining of gold and other metals in Curry County, but also aggregate mining in the area. 102 ppgs, 11.99

Origin of the Gold Bearing Black Sands of the Coast of South West Oregon - First released in 1943 and now out of print, this special reprint edition of "The Origin of the Black Sands of the South West Oregon Coast" was originally published by the Oregon Department of Geology and Mineral Industries. The text features hard to find information on the origin of the gold bearing black sands of the South West Oregon Coast, offering a unique insight to prospectors in search of Oregon's legendary beach gold. 52 ppgs, 8.99

South West Oregon Mining - Leading mining historian Kerby Jackson introduces us to six classic small mining publications on the Gold Mining Industry in Southern Oregon. This small book consists of a compilation of USGS J.S. Diller's "Mines of the Riddles Quadrangle", "The Rogue River Valley Coal Fields" and "Mineral Resources of the Grants Pass Quadrangle", the Grants Pass Commercial Club's rare publication "Mining in Josephine County, Oregon" and the USGS publication "The Distribution of Placer Gold in the Sixes River, South West Oregon". Also included is F.W. Libbey's legendary article on the Southern Oregon Mining Industry, "Lest We Forget", which appeared in the publication of the Oregon State Department of Geology and Mineral Industries in the early 1960's. This compilation offers a unique perspective on mining in South West Oregon and includes considerable information on mines in Josephine, Jackson and Coos Counties. 142 ppgs, 14.99

Geology and Mineral Resources of the Gasquet Quadrangle of California-Oregon - First published in 1953, it has been unavailable for over a century and sheds important light on the geological features and mineral resources of this portion of Northern California and Southern Oregon. 80 ppgs, 9.99

Idaho Mining Books

Gold in Idaho - Unavailable since the 1940's, this publication was originally compiled by the Idaho Bureau of Mines and includes details on gold mining in Idaho. Included is not only raw data on gold production in Idaho, but also valuable insight into where gold may be found in Idaho, as well as practical information on the gold bearing rocks and other geological features that will assist those looking for placer and lode gold in the State of Idaho. This volume also includes thirteen gold maps that greatly enhance the practical usability of the information contained in this small book detailing where to find gold in Idaho. **8.5" X 11", 72 ppgs. Retail Price: $9.99**

Geology of the Couer D'Alene Mining District of Idaho - Unavailable since 1961, this publication was originally compiled by the Idaho Bureau of Mines and Geology and includes details on the mining of gold, silver and other minerals in the famous Coeur D'Alene Mining District in Northern Idaho. Included are details on the early history of the Coeur D'Alene Mining District, local tectonic settings, ore deposit features, information on the mineral belts of the Osburn Fault, as well as detailed information on the famous Bunker Hill Mine, the Dayrock Mine, Galena Mine, Lucky Friday Mine and the infamous Sunshine Mine. This volume also includes sixteen hard to find maps. **8.5" X 11", 70 ppgs. Retail Price: $9.99**

The Gold Camps and Silver Cities of Idaho - Originally published in 1963, this important publication on Idaho Mining has not been available for nearly fifty years. Included are rare insights into the history of Idaho's Gold Rush, as well as the mad craze for silver in the Idaho Panhandle. Documented in fine detail are the early mining excitements at Boise Basin, at South Boise, in the Owyhees, at Deadwood, Long Valley, Stanley Basin and Robinson Bar, at Atlanta, on the famous Boise River, Volcano, Little Smokey, Banner, Boise Ridge, Hailey, Leesburg, Lemhi, Pearl, at South Mountain, Shoup and Ulysses, Yellow Jacket and Loon Creek. The story follows with the appearance of Chinese miners at the new mining camps on the Snake River, Black Pine, Yankee Fork, Bay Horse, Clayton, Heath, Seven Devils, Gibbonsville, Vienna and Sawtooth City. Also included are special sections on the Idaho Lead and Silver mines of the late 1800's, as well as the mining discoveries of the early 1900's that paved the way for Idaho's modern mining and mineral industry. Lavishly illustrated with rare historic photos, this volume provides a one of a kind documentary into Idaho's mining history that is sure to be enjoyed by not only modern miners and prospectors who still scour the hills in search of nature's treasures, but also those enjoy history and tromping through overgrown ghost towns and long abandoned mining camps. **8.5" X 11", 186 ppgs. Retail Price: $14.99**

Ore Deposits and Mining in North Western Custer County Idaho - Unavailable since 1913, this important publication was originally published by the Us Department of the Interior and has been unavailable for a century. Included are fine details on the geology, geography, gold placers and gold and silver bearing quartz veins of the mining region of North West Custer County, Idaho. Of particular interest is a rare look at the mines and prospects of the region, including those such as the Ramshorn Mine, SkyLark, Riverview, Excelsior, Beardsley, Pacific, Hoosier, Silver Brick, Forest Rose and dozens of others in the Bay Horse Mining District. Also covered are the mines of the Yankee Fork District such as the Lucky Boy, Badger, Black, Enterprise, Charles Dickens, Morrison, Golden Sunbeam, Montana, Golden Gate and others, as well as those in the Loon Mining District. **8.5" X 11", 126 ppgs. Retail Price: $12.99**

Gold Rush To Idaho - Unavailable since 1963, this important publication was originally published by the Idaho Bureau of Mines and has been unavailable for 50 years. "Gold Rush To Idaho" revisits the earliest years of the discovery of gold in Idaho Territory and introduces us to the conditions that the pioneer gold seekers met when they blazed a trail through the wilderness of Idaho's mountains and discovered the precious yellow metal at Oro Fino and Pierce. Subsequent rushes followed at places like Elk City, Newsome, Clearwater Station, Florence, Warrens and elsewhere. Of particular interest is a rare look at the hardships that the first miners in Idaho met with during their day to day existences and their attempts to bring law and order to their mining camps. **8.5" X 11", 88 ppgs. Retail Price: $9.99**

The Geology and Mines of Northern Idaho and North Western Montana - Unavailable since 1909, this important publication was originally published by the Us Department of the Interior and has been unavailable for a century. Included are fine details on the geology and geography of the mining regions of Northern Idaho and North Western Montana. Of particular interest is a rare look at the mines and prospects of the region, including those in the Pine Creek Mining District, Lake Pend Oreille district, Troy Mining District, Sylvanite District, Cabinet Mining District, Prospect Mining District and the Missoula Valley. Some of the mines featured include the Iron Mountain, Silver Butte, Snowshoe, Grouse Mountain Mine and others. **8.5" X 11", 142 ppgs. Retail Price: $12.99**

Mining in the Alturas Quadrangle of Blaine County Idaho - Unavailable since 1922, this important publication was originally published by the Idaho Bureau of Mines and has been unavailable for ninety years. Topics include the geology, rock formations and the formation of ore deposits in this important mining area of Idaho. Of particular focus is information on the local geology, quartz veins and ore deposits of this portion of Idaho. Included are hard to find details, including the descriptions and locations of numerous gold and silver mines in the area including the Silver King, Pilgrim, Columbia, Lone Jack, Sunbeam, Pride of the West, Lucky Boy, Scotia, Atlanta, Beaver-Bidwell and others mines and prospects. **8.5" X 11", 56 ppgs. Retail Price: $8.99**

Mining in Lemhi County Idaho - Originally published in 1913, this important book on Idaho Mining has not been available to miners for over a century. Included are rare insights into hundreds of gold, silver, copper and other mines in this famous Idaho mining area. Details include the locations, geology, history, production and other facts of the mines of this region, not only gold and silver hardrock mines, but also gold placer mines, lead-silver deposits, copper mines, cobalt-nickel deposits, tungsten and tin mines . It is lavishly illustrated with hard to find photos of the period and rare mining maps. Some of the vicinities featured include the Nicholia Mining District, Spring Mountain District, Texas District, Blue Wing District, Junction District, McDevitt District, Pratt Creek, Eldorado District, Kirtley Creek, Carmen Creek, Gibbonsville, Indian Creek, Mineral Hill District, Mackinaw, Eureka District, Blackbird District, YellowJacket District, Gravel Range District, Junction District, Parker Mountain and other mining districts. 8.5" X 11", 226 ppgs. Retail Price: $19.99

Mining in Shoshone County Idaho - First published in 1923, it has been unavailable for over a century and sheds important light on the mining history of Shoshone County, Idaho. Some of the topics include the history of mining in Shoshone County, a look at the local geology and ore characteristics of lead-silver deposits, zinc deposits, copper, antimony, gold and other minerals. Also included are insights into the history, production, characteristics and locations of numerous mines in the area. 198 ppgs, 15.99

Utah Mining Books

Fluorite in Utah - Unavailable since 1954, this publication was originally compiled by the USGS, State of Utah and U.S. Atomic Energy Commission and details the mining of fluorspar, also known as fluorite in the State of Utah. Included are details on the geology and history of fluorspar (fluorite) mining in Utah, including details on where this unique gem mineral may be found in the State of Utah. 8.5" X 11", 60 ppgs. Retail Price: $8.99

The Gold Hill Mining District of Utah - First published in 1935, it has been unavailable since those days and sheds important light on the mines, history and geology of Utah's Gold Hill Mining District. Included are rare insights into this important mining area, including the locations, histories and details of numerous mines. This volume is well illustrated with geological diagrams, as well as hard to find maps of some of the most important mines in this district. 202 ppgs., 19.99

The Mines, Miners and Minerals of Utah - First published in 1896, it has been unavailable since those days and sheds important light on the early mines and miners of Pioneer Utah, as well as the minerals which they won from the earth by laborious hard physical labor and sheer determination. Included are rare insights into the early mining history of Utah, as well details on hundreds of gold, silver and copper mines. 376 ppgs., 24.99

California Mining Books

The Tertiary Gravels of the Sierra Nevada of California - Mining historian Kerby Jackson introduces us to a classic mining work by Waldemar Lindgren in this important re-issue of The Tertiary Gravels of the Sierra Nevada of California. Unavailable since 1911, this publication includes details on the gold bearing ancient river channels of the famous Sierra Nevada region of California. 8.5" X 11", 282 ppgs. Retail Price: $19.99

The Mother Lode Mining Region of California - Unavailable since 1900, this publication includes details on the gold mines of California's famous Mother Lode gold mining area. Included are details on the geology, history and important gold mines of the region, as well as insights into historic mining methods, mine timbering, mining machinery, mining bell signals and other details on how these mines operated. Also included are insights into the gold mines of the California Mother Lode that were in operation during the first sixty years of California's mining history. 8.5" X 11", 176 ppgs. Retail Price: $14.99

Lode Gold of the Klamath Mountains of Northern California and South West Oregon - Unavailable since 1971, this publication was originally compiled by Preston E. Hotz and includes details on the lode mining districts of Oregon and California's Klamath Mountains. Included are details on the geology, history and important lode mines of the French Gulch, Deadwood, Whiskeytown, Shasta, Redding, Muletown, South Fork, Old Diggings, Dog Creek (Delta), Bully Choop (Indian Creek), Harrison Gulch, Hayfork, Minersville, Trinity Center, Canyon Creek, East Fork, New River, Denny, Liberty (Black Bear), Cecilville, Callahan, Yreka, Fort Jones and Happy Camp mining districts in California, as well as the Ashland, Rogue River, Applegate, Illinois River, Takilma, Greenback, Galice, Silver Peak, Myrtle Creek and Mule Creek districts of South Western Oregon. Also included are insights into the mineralization and other characteristics of this important mining region. 8.5" X 11", 100 ppgs. Retail Price: $10.99

Mines and Mineral Resources of Shasta County, Siskiyou County, Trinity County: California - Unavailable since 1915, this publication was originally compiled by the California State Mining Bureau and includes details on the gold mines of this area of Northern California. Also included are insights into the mineralization and other characteristics of this important mining region, as well as the location of historic gold mines. 8.5" X 11", 204 ppgs. Retail Price: $19.99

Geology of the Yreka Quadrangle, Siskiyou County, California - Unavailable since 1977, this publication was originally compiled by Preston E. Hotz and includes details on the geology of the Yreka Quadrangle of Siskiyou County, California. Also included are insights into the mineralization and other characteristics of this important mining region. 8.5" X 11", 78 ppgs. Retail Price: $7.99

Mines of San Diego and Imperial Counties, California - Originally published in 1914, this important publication on California Mining has not been available for a century. This publication includes important information on the early gold mines of San Diego and Imperial County, which were some of the first gold fields mined in California by early Spanish and Mexican miners before the 49ers came on the scene. Included are not only details on early mining methods in the area, production statistics and geological information, but also the location of the early gold mines that helped make California "The Golden State". Also included are details on the mining of other minerals such as silver, lead, zinc, manganese, tungsten, vanadium, asbestos, barite, borax, cement, clay, dolomite, fluospar, gem stones, graphite, marble, salines, petroleum, stronium, talc and others. 8.5" X 11", 116 ppgs. Retail Price: $12.99

Mines of Sierra County, California - Unavailable since 1920, this publication was originally compiled by the California State Mining Bureau and includes details on the gold mines of Sierra County, California. Also included are insights into the mineralization and other characteristics of this important mining region, as well as the location of historic gold mines. 8.5" X 11", 156 ppgs. Retail Price: $19.99

Mines of Plumas County, California - Unavailable since 1918, this publication was originally compiled by the California State Mining Bureau and includes details on the gold mines of Plumas County, California. Also included are insights into the mineralization and other characteristics of this important mining region, as well as the location of historic gold mines. 8.5" X 11", 200 ppgs. Retail Price: $19.99

Mines of El Dorado, Placer, Sacramento and Yuba Counties, California - Originally published in 1917, this important publication on California Mining has not been available for nearly a century. This publication includes important information on the early gold mines of El Dorado County, Placer County, Sacramento County and Yuba County, which were some of the first gold fields mined by the Forty-Niners during the California Gold Rush. Included are not only details on early mining methods in the area, production statistics and geological information, but also the location of the early gold mines that helped make California "The Golden State". Also included are insights into the early mining of chrome, copper and other minerals in this important mining area. 8.5" X 11", 204 ppgs. Retail Price: $19.99

Mines of Los Angeles, Orange and Riverside Counties, California - Originally published in 1917, this important publication on California Mining has not been available for nearly a century. This publication includes important information on the early gold mines of Los Angeles County, Orange County and Riverside County, which were some of the first gold fields mined in California by early Spanish and Mexican miners before the 49ers came on the scene. Included are not only details on early mining methods in the area, production statistics and geological information, but also the location of the early gold mines that helped make California "The Golden State". 8.5" X 11", 146 ppgs. Retail Price: $12.99

Mines of San Bernadino and Tulare Counties, California - Originally published in 1917, this important publication on California Mining has not been available for nearly a century. This publication includes important information on the early gold mines of San Bernadino and Tulare County, which were some of the first gold fields mined in California by early Spanish and Mexican miners before the 49ers came on the scene. Included are not only details on early mining methods in the area, production statistics and geological information, but also the location of the early gold mines that helped make California "The Golden State". Also included are details on the mining of other minerals such as copper, iron, lead, zinc, manganese, tungsten, vanadium, asbestos, barite, borax, cement, clay, dolomite, fluospar, gem stones, graphite, marble, salines, petroleum, stronium, talc and others. 8.5" X 11", 200 ppgs. Retail Price: $19.99

Chromite Mining in The Klamath Mountains of California and Oregon - Unavailable since 1919, this publication was originally compiled by J.S. Diller of the United States Department of Geological Survey and includes details on the chromite mines of this area of Northern California and Southern Oregon. Also included are insights into the mineralization and other characteristics of this important mining region, as well as the location of historic mines. Also included are insights into chromite mining in Eastern Oregon and Montana. 8.5" X 11", 98 ppgs. Retail Price: $9.99

Mines and Mining in Amador, Calaveras and Tuolumne Counties, California - Unavailable since 1915, this publication was originally compiled by William Tucker and includes details on the mines and mineral resources of this important California mining area. Included are details on the geology, history and important gold mines of the region, as well as insights into other local mineral resources such as asbestos, clay, copper, talc, limestone and others. Also included are insights into the mineralization and other characteristics of this important portion of California's Mother Lode mining region. 8.5" X 11", 198 ppgs. Retail Price: $14.99

The Cerro Gordo Mining District of Inyo County California - Unavailable since 1963, this publication was originally compiled by the United States Department of Interior. Included are insights into the mineralization and other characteristics of this important mining region of Southern California. Topics include the mining of gold and silver in this important mining district in Inyo County, California, including details on the history, production and locations of the Cerro Gordo Mine, the Morning Star Mine, Estelle Tunnel, Charles Lease Tunnel, Ignacio, Hart, Crosscut Tunnel, Sunset, Upper Newtown, Newtown, Ella, Perseverance, Newsboy, Belmont and other silver and gold mines in the Cerro Gordo Mining District. This volume also includes important insights into the fossil record, geologic formations, faults and other aspects of economic geology in this California mining district. 8.5" X 11", 104 ppgs. Retail Price: $10.99

Mining in Butte, Lassen, Modoc, Sutter and Tehama Counties of California - Unavailable since 1917, this publication was originally compiled by the United States Department of Interior. Included are insights into the mineralization and other characteristics of this important mining region of California. Topics include the mining of asbestos, chromite, gold, diamonds and manganese in Butte County, the mining of gold and copper in the Hayden Hill and Diamond Mountain mining districts of Lassen County, the mining of coal, salt, copper and gold in the High Grade and Winters mining districts of Modoc County, gold mining in Sutter County and the mining of gold, chromite, manganese and copper in Tehama County. This volume also includes the production records and locations of numerous mines in this important mining region. 8.5" X 11", 114 ppgs. Retail Price: $11.99

Mines of Trinity County California - Originally published in 1965, this important publication on California Mining has not been available for nearly fifty years. This publication includes important information on mines and mining in Trinity County, California, as well insights into the mineralization and geology of this important mining area in Northern California. Included are extensive details on hardrock and placer gold mines and prospects, including charts showing the locations of these historic mines.. 8.5" X 11", 144 ppgs. Retail Price: $12.99

Mines of Kern County California - Originally published in 1962, this important publication on California Mining has not been available for nearly fifty years. This publication includes important information on mines and mining in Kern County, California, as well insights into the mineralization and geology of this important mining area in California. Included are extensive details on hardrock and placer gold mines and prospects, including charts showing the locations of these historic mines. 8.5" X 11", 398 ppgs. Retail Price: $24.99

Mines of Calaveras County California - Originally published in 1962, this important publication on California Mining has not been available for nearly fifty years. This publication includes important information on mines and mining in Calaveras County, California, as well insights into the mineralization and geology of this important mining area in Northern California. Included are extensive details on hardrock and placer gold mines and prospects, including charts showing the locations of these historic mines. 8.5" X 11", 236 ppgs. Retail Price: $19.99

Lode Gold Mining in Grass Valley California - Unavailable since 1940, this publication was originally compiled by the United States Department of Interior. Included are insights into the gold mineralization and other characteristics of this important mining region of Nevada County, California. This volume also includes important insights into the geologic formations, faults and other aspects of economic geology in this California mining district. Of particular interest are the fine details on many hardrock gold mines in the area, including their locations, histories, development and mineralization. Some of the mines featured include the Gold Hill Mine, Massachusetts Hill, Boundary, Peabody, Golden Center, North Star, Omaha, Lone Jack, Homeward Bound, Hartery, Wisconsin, Allison Ranch, Phoenix, Kate Hayes, W.Y.O.D., Empire, Rich Hill, Daisy Hill, Orleans, Sultana, Centennial, Conlin, Ben Franklin, Crown Point and many others. 8.5" X 11", 148 ppgs. Retail Price: $12.99

Lode Mining in the Alleghany District of Sierra County California - Unavailable since 1913, this publication was originally compiled by the United States Department of Interior. Included are insights into the mineralization and other characteristics of this important mining region of Sierra County. Included are details on the history, production and locations of numerous hardrock gold mines in this famous California area, including the Tightner Mine, Minnie D., Osceola, Eldorado, Twenty One, Sherman, Kenton, Oriental, Rainbow, Plumbago, Irelan, Gold Canyon, North Fork, Federal, Kate Hardy and others. This volume also includes important insights into the fossil record, geologic formations, faults and other aspects of economic geology in this California mining district. 8.5" X 11", 48 ppgs. Retail Price: $7.99

Six Months In The Gold Mines During The California Gold Rush - Unavailable since 1850, this important work is a first hand account of one "49'ers" personal experience during the great California Gold Rush, shedding important light on one of the most exciting periods in the history of not only California, but also the world. Compiled from journals written between 1847 and 1849 by E. Gould Buffum, a native of New York, "Six Months In The Gold Mines During The California Gold Rush" offers a rare look into the day to day lives of the people who came to California to work in her gold mines when the state was still a great frontier. 8.5" X 11", 290 ppgs. Retail Price: $19.99

Quartz Mines of the Grass Valley Mining District of California - Unavailable since 1867, this important publication has not been available since those days. This rare publication offers a short dissertation on the early hardrock mines in this important mining district in the California Mother Lode region between the 1850's and 1860's. Also included are hard to find details on the mineralization and locations of these mines, as well as how they were operated in those day. **8.5" X 11", 44 ppgs. Retail Price: $8.99**

Gold Rush on the Feather River - First published in 1924, this short publication by G.C. Mansfield sheds important light on the early history of gold mining on the Feather River. Included are rare insights into the first decade of gold mining and the early mining camps of the Feather River during the 1850's. 64 ppgs., 9.99

The Bodie Mining District of California - First published in 1986, it has been unavailable since those days and sheds important light on this famous mining area. Included are the history, characteristics and locations of numerous old mines around the ghost town of Bodie.
64 ppgs, 8.99

Geology and Mineral Resources of the Gasquet Quadrangle of California-Oregon - First published in 1953, it has been unavailable for over a century and sheds important light on the geological features and mineral resources of this portion of Northern California and Southern Oregon.
80 ppgs, 9.99

Alaska Mining Books

Ore Deposits of the Willow Creek Mining District, Alaska - Unavailable since 1954, this hard to find publication includes valuable insights into the Willow Creek Mining District near Hatcher Pass in Alaska. The publication includes insights into the history, geology and locations of the well known mines in the area, including the Gold Cord, Independence, Fern, Mabel, Lonesome, Snowbird, Schroff-O'Neil, High Grade, Marion Twin, Thorpe, Webfoot, Kelly-Willow, Lane, Holland and others. **8.5" X 11", 96 ppgs. Retail Price: $9.99**

The Juneau Gold Belt of Alaska - Unavailable since 1906, this hard to find publication includes valuable insights into the gold mines around Juneau, Alaska. The publication includes important details into the history, geology and locations of the well known gold mines and prospects in the area, including those around Windham Bay, Holkham Bay, Port Snettisham, on Grindstone and Rhine Creeks, Gold Creek, Douglas Island, Salmon Creek, Lemon Creek, Nugget Creek, from the Mendenhall River to Berners Bay, McGinnis Creek, Montana Creek, Peterson Creek, Windfall Creek, the Eagle River, Yankee Basin, Yankee Curve, Kowee Creek and elsewhere. Not only are gold placer mines included, but also hardrock gold mines. **8.5" X 11", 224 ppgs. Retail Price: $19.99**

Mining in the Jumbo Basin of Alaska - Unavailable since 1953, this hard to find publication includes valuable insights into the mines and geology of the Jumbo Basin. The publication includes important details into the history, geology and locations of the well known gold mines and prospects in the famous Jumbo Basin Mining Region of Alaska.
72 ppgs, 9.99

The Rampart Placer Gold Region of Alaska - Unavailable since 1906, this hard to find publication includes valuable insights into the placer gold mines of the Rampart Mining Region. The publication includes important details into the history, geology and locations of the well known gold mines and prospects in the famous Rampart Mining Region of Alaska.
78 ppgs, 10.99

Arizona Mining Books

Mines and Mining in Northern Yuma County Arizona - Originally published in 1911, this important publication on Arizona Mining has not been available for over a hundred years. Included are rare insights into the gold, silver, copper and quicksilver mines of Yuma County, Arizona together with hard to find maps and photographs. Some of the mines and mining districts featured include the Planet Copper Mine, Mineral Hill, the Clara Consolidated Mine, Viati Mine, Copper Basin prospect, Bowman Mine, Quartz King, Billy Mack, Carnation, the Wardwell and Osbourne, Valensuella Copper, the Mariquita, Colonial Mine, the French American, the New York-Plomosa, Guadalupe, Lead Camp, Mudersbach Copper Camp, Yellow Bird, the Arizona Northern (Salome Strike), Bonanza (Harqua Hala), Golden Eagle, Hercules, Socorro and others. **8.5" X 11", 144 ppgs. Retail Price: $11.99**

The Aravaipa and Stanley Mining Districts of Graham County Arizona - Originally published in 1925, this important publication on Arizona Mining has not been available for nearly ninety years. Included are rare insights into the gold and silver mines of these two important mining districts, together with hard to find maps. **8.5" X 11", 140 ppgs. Retail Price: $11.99**

Gold in the Gold Basin and Lost Basin Mining Districts of Mohave County, Arizona - This volume contains rare insights into the geology and gold mineralization of the Gold Basin and Lost Basin Mining Districts of Mohave County, Arizona that will be of benefit to miners and prospectors. Also included is a significant body of information on the gold mines and prospects of this portion of Arizona. This volume is lavishly illustrated with rare photos and mining maps. **8.5" X 11", 188 ppgs. Retail Price: $19.99**

Mines of the Jerome and Bradshaw Mountains of Arizona - This important publication on Arizona Mining has not been available for ninety years. This volume contains rare insights into the geology and ore deposits of the Jerome and Bradshaw Mountains of Arizona that will be of benefit to miners and prospectors who work those areas. Included is a significant body of information on the mines and prospects of the Verde, Black Hills, Cherry Creek, Prescott, Walker, Groom Creek, Hassayampa, Bigbug, Turkey Creek, Agua Fria, Black Canyon, Peck, Tiger, Pine Grove, Bradshaw, Tintop, Humbug and Castle Creek Mining Districts. This volume is lavishly illustrated with rare photos and mining maps. **8.5" X 11", 218 ppgs. Retail Price: $19.99**

The Ajo Mining District of Pima County Arizona - This important publication on Arizona Mining has not been available for nearly seventy years. This volume contains rare insights into the geology and mineralization of the Ajo Mining District in Pima County, Arizona and in particular the famous New Cornelia Mine. **8.5" X 11", 126 ppgs. Retail Price: $11.99**

Mining in the Santa Rita and Patagonia Mountains of Arizona - Originally published in 1915, this important publication on Arizona Mining has not been available for nearly a century. Included are rare insights into hundreds of gold, silver, copper and other mines in this famous Arizona mining area. Details include the locations, geology, history, production and other facts of the mines of this region. **8.5" X 11", 394 ppgs. Retail Price: $24.99**

Mining in the Bisbee Quadrangle of Arizona - Originally published in 1906, this important publication on Arizona Mining has not been available for nearly a century. Included are rare insights into hundreds of gold, silver, copper and other mines in this famous Arizona mining area. Details include the locations, geology, history, production and other facts of the mines of this important mining region. **8.5" X 11", 188 ppgs. Retail Price: $14.99**

Placer Gold Mining in Arizona - Unavailable since 1922, this hard to find publication includes valuable insights into the placer gold mines of the Arizona. Originally released as "Placer Gold of Arizona", despite its small size, this publication includes important details into the history, geology and locations of the well known placer gold mines and prospects in the State of Arizona. **48 ppgs, 8.99**

Gold and Copper Mining near Payson, Arizona - Written in 1915, this hard to find publication includes valuable insights into the gold and copper mining industry of Arizona. Highlighted here are the gold and copper mines near Payson, Arizona. **68 ppgs, 8.99**

Lode Gold Mining in Arizona - Unavailable since 1934, this hard to find publication, originally released as "Arizona Lode Gold Mines and Gold Mining" includes valuable insights into the gold mining industry of Arizona. Included are valuable insights into over 150 hardrock gold mines in over 30 different mining districts in Arizona. **278 ppgs, 21.99**

Mining in the Dragoon Quadrangle of Cochise County, Arizona - Unavailable since 1964, this hard to find publication includes valuable insights into the mines of the Dragoon Quadrangle Mining Region. The publication includes important details into the history, geology and locations of the well known mines and prospects in this famous mining region of Arizona. **224 ppgs., 19.99**

Directory of Operating Mines in Arizona in 1915 - Unavailable since 1916, this hard to find publication includes valuable insights into the mines of Arizona. This small publication includes a complete list of the mines that were operating in the State of Arizona during 1915 and includes details such as general location, owners and some basic facts about each mining operation. **52 ppgs. 8.99**

Arizona Ore Deposits - Unavailable since 1938, this hard to find publication includes valuable insights into some ore deposits of Arizona. Included are valuable insights into the formation and characteristics of valuable ore deposits in the Jerome, Miami, Inspiration, Clifton, Morenci, Ray, Ajo, Eureka, Tombstone and Magma mining districts. Included are details into some of the major gold, silver and copper mines of these important Arizona mining areas. **160 ppgs, 14.99**

Montana Mining Books

A History of Butte Montana: The World's Greatest Mining Camp - First published in 1900 by H.C. Freeman, this important publication sheds a bright light on one of the most important mining areas in the history of The West. Together with his insights, as well as rare photographs of the periods, Harry Freeman describes Butte and its vicinity from its early beginnings, right up to its flush years when copper flowed from its mines like a river. At the time of publication, Butte, Montana was known worldwide as "The Richest Mining Spot On Earth" and produced not only vast amounts of copper, but also silver, gold and other metals from its mines. Freeman illustrates, with great detail, the most important mines in the vicinity of Butte, providing rare details on their owners, their history and most importantly, how the mines operated and how their treasures were extracted. Of particular interest are the dozens of rare photographs that depict mines such as the famous Anaconda, the Silver Bow, the Smoke House, Moose, Paulin, Buffalo, Little Minah, the Mountain Consolidated, West Greyrock, Cora, the Green Mountain, Diamond, Bell, Parnell, the Neversweat, Nipper, Original and many others. 8.5" X 11", 142 ppgs. Retail Price: $12.99

The Butte Mining District of Montana - This important publication on Montana Mining has not been available for over a century. Included are rare insights into the gold, copper and silver mines of Butte, Montana together with hard to find maps and photographs. Some of the topics include the early history of gold, silver and copper mining in the Butte area, insight into the geology of its mining areas, the local distribution of gold, silver and copper ores, as well their composition and how to identify them. Also included are detailed facts about the mines in the Butte Mining District, including the famous Anaconda Mine, Gagnon, Parrot, Blue Vein, Moscow, Poulin, Stella, Buffalo, Green Mountain, Wake Up Jim, the Diamond-Bell Group, Mountain Consolidated, East Greyrock, West Greyrock, Snowball, Corra, Speculator, Adirondack, Miners Union, the Jessie-Edith May Group, Otisco, Iduna, Colorado, Lizzie, Cambers, Anderson, Hesperus, Preferencia and dozens of others. 8.5" X 11", 298 ppgs. Retail Price: $24.99

Mines of the Helena Mining Region of Montana - This important publication on Montana Mining has not been available for over a century. Included are rare insights into the gold, copper and silver mines of the vicinity of Helena, Montana, including the Marysville Mining District, Elliston Mining District, Rimini Mining District, Helena Mining District, Clancy Mining District, Wickes Mining District, Boulder and Basin Mining Districts and the Elkhorn Mining District. Some of the topics include the early history of gold, silver and copper mining in the Helena area, insight into the geology of its mining areas, the local distribution of gold, silver and copper ores, as well their composition and how to identify them. Also included are detailed facts, history, geology and locations of over one hundred gold, silver and copper mines in the area . 8.5" X 11", 162 ppgs, Retail Price: $14.99

Mines and Geology of the Garnet Range of Montana - This important publication on Montana Mining has not been available for over a century. Included are rare insights into the gold, copper and silver mines of the vicinity of this important mining area of Montana. Some of the topics include the early history of gold, silver and copper mining in the Garnet Mountains, insight into the geology of its mining areas, the local distribution of gold, silver and copper ores, as well their composition and how to identify them. Also included are detailed facts, history, geology and locations of numerous gold, silver and copper mines in the area . 8.5" X 11", 100 ppgs, Retail Price: $11.99

Mines and Geology of the Philipsburg Quadrangle of Montana - This important publication on Montana Mining has not been available for over a century. Included are rare insights into the gold, copper and silver mines of the vicinity of this important mining area of Montana. Some of the topics include the early history of gold, silver and copper mining in the Philipsburg Quadrangle, insight into the geology of its mining areas, the local distribution of gold, silver and copper ores, as well their composition and how to identify them. Also included are detailed facts, history, geology and locations of over one hundred gold, silver and copper mines in the area 8.5" X 11", 290 ppgs, Retail Price: $24.99

Geology of the Marysville Mining District of Montana - Included are rare insights into the mining geology of the Marysville Mining District. Some of the topics include the early history of gold, silver and copper mining in the area, insight into the geology of its mining areas, the local distribution of gold, silver and copper ores, as well their composition and how to identify them. Also included are detailed facts, history, geology and locations of gold, silver and copper mines in the area 8.5" X 11", 198 ppgs, Retail Price: $19.99

The Geology and Mines of Northern Idaho and North Western Montana- See listing under Idaho.

The History of Gold Dredging in Montana - Unavailable since 1916, this important publication was originally published by the Us Bureau of Mines and has been unavailable for a century. A century and more ago, giant dredging machines dug in Montana's rivers and creeks in search of illusive golden riches. First appearing in California in the 1850's, gold dredges finally reached their peak of development in Siberia and New Zealand before becoming popular again in the United States. This book offers a unique historical perspective on the gold dredges that once operated in Montana. This book on Montana mining history is lavishly illustrated with dozens of rare historic photos gold dredges that once operated in Montana, as well as hard to locate plans on how these dredges were designed. 120 ppgs., 11.99

Nevada Mining Books

The Bull Frog Mining District of Nevada - Unavailable since 1910, this publication was originally compiled by the United States Department of Interior. This volume also includes important insights into the geologic formations, faults and other aspects of economic geology in this Nevada mining district. Of particular interest are the fine details on many mines in the area, including their locations, histories, development and mineralization. Some of the mines featured include the National Bank Mine, Providence, Gibraltor, Tramps, Denver, Original Bullfrog, Gold Bar, Mayflower, Homestake-King and other mines and prospects. 8.5" X 11", 152 ppgs, Retail Price: $14.99

History of the Comstock Lode - Unavailable since 1876, this publication was originally released by John Wiley & Sons. This volume also includes important insights into the famous Comstock Lode of Nevada that represented the first major silver discovery in the United States. During its spectacular run, the Comstock produced over 192 million ounces of silver and 8.2 million ounces of gold. Not only did the Comstock result in one of the largest mining rushes in history and yield immense fortunes for its owners, but it made important contributions to the development of the State of Nevada, as well as neighboring California. Included here are important details on not only the early development and history of the Comstock, but also rare early insight into its mines, ore and its geology.8.5" X 11", 244 ppgs, Retail Price: $19.99

The Pioche Mining District of Nevada - First published in 1932, it has been unavailable for over a century and sheds important light on the mining history of Nevada. Some of the topics include the history of mining in this district, as well as the characteristics of its mineral and ore deposits. Also included are insights into the history, production, characteristics and locations of numerous mines in the area. Some of the mines include the Combined Metals, Pioche, Ely Valley, No. 10, Poorman, Wide Awake, Alps, Prince, Virginia Louise, Half Moon, Abe Lincoln, Fairview, Bristol Silver, National, Vesuvius, Inman, Tempest, Hillside, Jackrabbit, Lucky Star, Fortuna, Mendha, Manhattan, Hamburg, Comet, Lyndon and others. 108 ppgs 10.99

The Yerington Mining District of Nevada - First published in 1932, it has been unavailable for over a century and sheds important light on the mining history of Nevada. Some of the topics include the history of mining in this district, as well as the characteristics of its mineral and ore deposits. Also included are insights into the history, production, characteristics and locations of numerous mines in the area. Some of the mines include the Bluestone, Mason Valley, Malachite, McConnell, Greenwood, Western Nevada, Ludwig, Douglas Hill, Casting Copper, Montana-Yerington, Empire, Jim Beatty, Terry and McFarland, Blue Jay and others. 92 ppgs, 10.99

The Genesis of the Ores of Tonopah Nevada - Unavailable since 1918, this hard to find publication includes valuable insights into the gold mines around Tonopah, Nevada. The publication includes important details into the geology of mines in the Tonopah Mining District of Nevada. 90 ppgs, 10.99

Mining Camps of Elko, Lander and Eureka Counties Nevada - Unavailable since 1910, this hard to find publication includes valuable insights into the mining camps of Elko, Lander and Eureka Counties, Nevada. The publication includes important details into the history of mines and mining in these three Nevada counties. 154 ppgs, 12.99

Ore Deposits of the Bullfrog Quadrangle - Unavailable since 1964 and released as "Geology of Bullfrog Quadrangle and Ore Deposits Related to Bullfrog Hills Caldera, Nye County, Nevada and Inyo County, California". The publication includes important details into the geology of mines in the Bullfrog Quadrangle of Nye County, Nevada and Inyo County, California. 52 ppgs, 9.99

Mining in Eureka County Nevada - Unavailable since 1879, this hard to find publication includes valuable insights into the early mining history off Eureka County, Nevada. The publication includes important details into the early history of the mines of Eureka County, as well as their development, production and how their ores were treated. Also included are details on the 1872 Mining Act, as well as the local rules, regulations and customs of the miners in Eureka County.134 ppgs, 12.99

Colorado Mining Books

Ores of The Leadville Mining District - Unavailable since 1926, this publication was originally compiled by the United States Department of Interior. This volume also includes important insights into the ores and mineralization of the Leadville Mining District in Colorado. Topics include historic ore prospecting methods, local geology, insights into ore veins and stockworks, the local trend and distribution of ore channels, reverse faults, shattered rock above replacement ore bodies, mineral enrichment in oxidized and sulphide zones and more. 8.5" X 11", 66 ppgs, **Retail Price: $8.99**

Mining in Colorado - Unavailable since 1926, this publication was originally compiled by the United States Department of Interior. This volume also includes important insights into the mining history of Colorado from its early beginnings in the 1850's right up to the mid 1920's. Not only is Colorado's gold mining heritage included, but also its silver, copper, lead and zinc mining industry. Each mining area is treated separately, detailing the development of Colorado's mines on a county by county basis. 8.5" X 11", 284 ppgs, **Retail Price: $19.99**

Gold Mining in Gilpin County Colorado - Unavailable since 1876, this publication was originally compiled by the Register Steam Printing House of Central City, Colorado. A rare glimpse at the gold mining history and early mines of Gilpin County, Colorado from their first discovery in the 1850's up to the "flush years" of the mid 1870's. Of particular interest is the history of the discovery of gold in Gilpin County and details about the men who made those first strikes. Special focus is given to the early gold mines and first mining districts of the area, many of which are not detailed in other books on Colorado's gold mining history. 8.5" X 11", 156 ppgs, **Retail Price: $12.99**

Mining in the Gold Brick Mining District of Colorado - Important insights into the history of the Gold Brick Mining District, as well as its local geography and economic geology. Also included are the histories and locations of historic mines in this important Colorado Mining District, including the Cortland, Carter, Raymond, Gold Links, Sacramento, Bassick, Sandy Hook, Chronicle, Grand Prize, Chloride, Granite Mountain, Lucille, Gray Mountain, Hilltop, Maggie Mitchell, Silver Islet, Revenue, Roosevelt, Carbonate King and others. In addition to hardrock mining, are also included are details on gold placer mining in this portion of Colorado. 8.5" X 11", 140 ppgs, **Retail Price: $12.99**

Ore Deposits of the London Fault of Colorado - First published in 1941, it has been unavailable since those days and sheds important light on the mines and mineral deposits of the London Fault in Central Colorado's Alma Mining District. This publication sheds important light on the gold veins and lead-silver deposits of the Alma Mining District. Included are geologic details on the London Mine, American Mine, Havigorst Tunnel, Ophir Mine, Mosher Tunnel, London-Butte Mine, Venture Shaft, Hard-To-Beat Mine, Oliver Twist Tunnel, Sacramento Mine, Mudsill Mine, Sherwood Mine, Wagner, Barcoe Tunnel and other mines in this important mining region. 110 ppgs., 10.99

The Mines of Colorado - First published in 1867, it has been unavailable since those days and sheds important light on Colorado's early mining history. Written shortly after the events took place, this publication sheds important light on the Pike's Peak Gold Rush, the discovery of gold on Ralston Creek and Dry Creek in the 1850's, as well as details on the first wave of miners into Colorado and their trials and tribulations as they crossed the Great Plains. Also included are details on early discoveries of lode gold in the mountainous regions of Colorado, details on the early mines hardrock and placer mines, and much more. It is a veritable treasure trove on Colorado's early mining history and will be of great importance to anyone who is interested in the mining of gold or other minerals in Colorado, as well as those interested in the history of the state. 478 ppgs., 29.99

The La Plata Mining District of Colorado - Originally titled "Geology and Ore Deposits in the Vicinity of the La Plata District of Colorado" and first published in 1949, it has been unavailable since those days and sheds important light on the mines and mineral deposits of the La Plata Mining District of Colorado. 214 ppgs., 19.99

Washington Mining Books

The Republic Mining District of Washington - Unavailable since 1910, this important publication was originally published by the Washington Geologic Survey and has been unavailable for a century. Topics include the geology, rock formations and the formation of ore deposits in this important mining area of Washington State. Also included are hard to find details on the geology, history and locations of dozens of mines in the area. Some of the mines featured include the New Republic Mine, Ben Hur, Morning Glory, the South Republic Mine, Quilp, Surprise, Black Tail, Lone Pine, San Poil, Mountain Lion, Tom Thumb, Elcaliph and many others. **8.5" X 11", 94 ppgs, Retail Price: $10.99**

The Myers Creek and Nighthawk Mining Districts of Washington - Unavailable since 1911, this important publication was originally published by the Washington Geologic Survey and has been unavailable for a century. Topics include the geology, rock formations and the formation of ore deposits in these important mining areas of Washington State. Also included are hard to find details on the geology, history and locations of dozens of mines in the area. Some of the mines featured include the Grant Mine, Monterey, Nip and Tuck, Myers Creek, Number Nine, Neutral, Rainbow, Aztec, Crystal Butte, Apex, Butcher Boy, Molson, Mad River, Olentangy, Delate, Kelsey, Golden Chariot, Okanogan, Ohio, Forty-Ninth Parallel, Nighthawk, Favorite, Little Chopaka, Summit, Number One, California, Peerless, Caaba, Prize Group, Ruby, Mountain Sheep, Golden Zone, Rich Bar, Similkameen, Kimberly, Triune, Hiawatha, Trinity, Hornsilver, Maquae, Bellevue, Bullfrog, Palmer Lake, Ivanhoe, Copper World and many others. **8.5" X 11", 136 ppgs, Retail Price: $12.99**

The Blewett Mining District of Washington - Unavailable since 1911, this important publication was originally published by the Washington Geologic Survey and has been unavailable for a century. Topics include the geology, rock formations and the formation of ore deposits in this important mining area of Washington State. Also included are hard to find details on the geology, history and locations of dozens of mines in the area. Some of the mines featured include the Washington Meteor, Alta Vista, Pole Pick, Blinn, North Star, Golden Eagle, Tip Top, Wilder, Golden Guinea, Lucky Queen, Blue Bell, Prospect, Homestake, Lone Rock, Johnson, and others. **8.5" X 11", 134 ppgs, Retail Price: $12.99**

Silver Mining In Washington - Unavailable since 1955, this important publication was originally published by the Washington Geologic Survey. Featured are the hard to find locations and details pertaining to Washington's silver mines. **8.5" X 11", 180 ppgs, Retail Price: $15.99**

The Mines of Snohomish County Washington - Unavailable since 1942, this important publication was originally published by the Washington Geologic Survey and has been unavailable for seventy years. Featured are details on a large number of gold, silver, copper, lead and other metallic mineral mines. Included are the locations of each historic mine, along with information on the commodity produced. **8.5" X 11", 98 ppgs, Retail Price: $10.99**

The Mines of Chelan County Washington - Unavailable since 1943, this important publication was originally published by the Washington Geologic Survey and has been unavailable for seventy years. Featured are details on a large number of gold, silver, copper, lead and other metallic mineral mines. Included are the locations of each historic mine, along with information on the commodity. **8.5" X 11", 88 ppgs, Retail Price: $9.99**

Metal Mines of Washington - Unavailable since 1921, this important publication was originally published by the Washington Geologic Survey and has been unavailable for nearly ninety years. Widely considered a masterpiece on the Washington Mining Industry, "Metal Mines of Washington" sheds light on the important details of Washington's early mining years. Featured are details on hundreds of gold, silver, copper, lead and other metallic mineral mines. Included are hard to find details on the mineral resources of this state, as well as the locations of historic mines. Lavishly illustrated with maps and historic photos and complete with a glossary to explain any technical terms found in the text, this is one of the most important works on mining in the State of Washington. No prospector or miner should be without it if they are interested in mining in Washington. **8.5" X 11", 396 ppgs, Retail Price: $24.99**

Gem Stones In Washington - Unavailable since 1949, this important publication was originally published by the Washington Geologic Survey and has been unavailable since first published. Included are details on where to find naturally occurring gem stones in the State of Washington, including quartz crystal, amethyst, smoky quartz, milky quartz, agates, bloodstone, carnelian, chert, flint, jasper, onyx, petrified wood, opal, fire opal, hyalite and others. **8.5" X 11", 54 ppgs, Retail Price: $8.99**

The Covada Mining District of Washington - Unavailable since 1913, this important publication was originally published by the Washington Geologic Survey and has been unavailable for a century. Topics include the geology, rock formations and the formation of ore deposits in this important mining area of Washington State. Also included are hard to find details on the geology, history and locations of dozens of mines in the area. Some of the mines featured include the Admiral, Advance, Algonkian, Big Bug, Big Chief, Big Joker, Black Hawk, Black Tail, Black Thorn, Captain, Cherokee Strip, Colorado, Dan Patch, Dead Shot, Etta, Good Ore, Greasy Run, Great Scott, Idora, IXL, Jay Bird, Kentucky Bell, King Solomon, Laurel, Laura S, Little Jay, Meteor, Neglected, Northern Light, Old Nell, Plymouth Rock, Polaris, Quandary, Reserve, Shoo Fly, Silver Plume, Three Pines, Vernie, White Rose and dozens of others. **8.5" X 11", 114 ppgs, Retail Price: $10.99**

The Index Mining District of Washington - Unavailable since 1912, this important publication was originally published by the Washington Geologic Survey and has been unavailable for a century. Topics include the geology, rock formations and the formation of ore deposits in this important mining area of Washington State. Also included are hard to find details on the geology, history and locations of dozens of mines in the area. Some of the mines featured include the Sunset, Non-Pareil, Ethel Consolidated, Kittaning, Merchant, Homestead, Co-operative, Lost Creek, Uncle Sam, Calumet, Florence-Rae, Bitter Creek, Index Peacock, Gunn Peak, Helena, North Star, Buckeye. Copper Bell, Red Cross and others. **8.5" X 11", 114 ppgs, Retail Price: $11.99**

Mining & Mineral Resources of Stevens County Washington - Unavailable since 1920, this important publication was originally published by the Washington Geologic Survey and has been unavailable for a century. Topics include the geology, rock formations and the formation of ore deposits in these important mining areas of Washington State. Also included are hard to find details on the geology, history and locations of hundreds of mines in the area. **8.5" X 11", 372 ppgs, Retail Price: $24.99**

The Mines and Geology of the Loomis Quadrangle Okanogan County, Washington - Unavailable since 1972, this important publication was originally published by the Washington Geologic Survey and has been unavailable for a century. Topics include the geology, rock formations and the formation of ore deposits in this important mining area of Washington State. Also included are hard to find details on the geology, history and locations of dozens of gold, copper, silver and other mines in the area. **8.5" X 11", 150 ppgs, Retail Price: $12.99**

The Conconully Mining District of Okanogan County Washington - Unavailable since 1973, this important publication was originally published by the Washington Geologic Survey and has been unavailable for a century. Topics include the geology, rock formations and the formation of ore deposits in this important mining area of Washington State, which also includes Salmon Creek, Blue Lake and Galena. Also included are hard to find details on the geology, mining history and locations of dozens of mines in the area. Some of the mines include Arlington, Fourth of July, Sonny Boy, First Thought, Last Chance, War Eagle-Peacock, Wheeler, Mohawk, Lone Star, Woo Loo Moo Loo, Keystone, Hughes, Plant-Callahan, Johnny Boy, Leuena, Gubser, John Arthur, Tough Nut, Homestake, Key and many others **8.5" X 11", 68 ppgs, Retail Price: $8.99**

Wyoming Mining Books

Mining in the Laramie Basin of Wyoming - Unavailable since 1909, this publication was originally compiled by the United States Department of Interior. Also included are insights into the mineralization and other characteristics of this important mining region, especially in regards to coal, limestone, gypsum, bentonite clay, cement, sand, clay and copper. **8.5" X 11", 104 ppgs, Retail Price: $11.99**

New Mexico Mining Books

The Mogollon Mining District of New Mexico - Unavailable since 1927, this important publication was originally published by the US Department of Interior and has been unavailable for 80 years. Topics include the geology, rock formations and the formation of ore deposits in this important mining area in New Mexico. Of particular focus is information on the history and production of the ore deposits in this area, their form and structure, vein filling, their paragenesis, origins and ore shoots, as well as oxidation and supergene enrichment. Also included are hard to find details, including the descriptions and locations of numerous gold, silver and other types of mines, including the Eureka, Pacific, South Alpine, Great Western, Enterprise, Buffalo, Mountain View, Floride, Gold Dust, Last Chance, Deadwood, Confidence, Maud S., Deep Down, Little Fanney, Trilby, Johnson, Alberta, Comet, Golden Eagle, Cooney, Queen, the Iron Crown, Eberle, Clifton, Andrew Jackson mine, Mascot and others. **8.5" X 11", 144 ppgs, Retail Price: $12.99**

The Percha Mining District of Kingston New Mexico - Unavailable since 1883, this important publication was originally published by the Kingston Tribune and has been unavailable for over one hundred and thirty five years. Having been written during the earliest years of gold and silver mining in the Percha Mining District, unlike other books on the subject, this work offers the unique perspective of having actually been written while the early mining history of this area was still being made. In fact, the work was written so early in the development of this area that many of the notable mines in the Percha District were less than a few years old and were still being operated by their original discoverers with the same enthusiasm as when they were first located. Included are hard to find details on the very earliest gold and silver mines of this important mining district near Kingston in Sierra County, New Mexico. **8.5" X 11", 68 ppgs, Retail Price: $9.99**

East Coast Mining Books

<u>The Gold Fields of the Southern Appalachians</u> - Unavailable since 1895, this important publication was originally published by the US Department of Interior and has been unavailable for nearly 120 years. Topics include the geology, rock formations and the formation of ore deposits in this important mining area of the American South. Of particular focus is information on the history and statistics of the ore deposits in this area, their form and structure and veins. Also included are details on the placer gold deposits of the region. The gold fields of the Georgian Belt, Carolinian Belt and the South Mountain Mining District of North Carolina are all treated in descriptive detail. Included are hard to find details, including the descriptions and locations of numerous gold mines in Georgia, North Carolina and elsewhere in the American South. Also included are details on the gold belts of the British Maritime Provinces and the Green Mountains. **8.5" X 11", 104 ppgs, Retail Price: $9.99**

Gold Rush Tales Series

Millions in Siskiyou County Gold - In this first volume of the "Gold Rush Tales" series, leading mining historian and editor Kerby Jackson, introduces us to the story of how millions of dollars worth of gold was discovered in Siskiyou County during the California Gold Rush. Lavishly illustrated with photos from the 19th Century, this hard to find information was first published in 1897 and sheds important light onto the gold rush era in Siskiyou County, California and the experiences of the men who dug for the gold and actually found it. **8.5" X 11", 82 ppgs, Retail Price: $9.99**

The California Rand in the Days of '49 - In this second volume of the "Gold Rush Tales" series, leading mining historian and editor Kerby Jackson, introduces us to four tales from the California Gold Rush. Lavishly illustrated with photos from the 19th Century, this hard to find information was first published in 1890's and includes the stories of "California's Rand", details about Chinese miners, how one early miner named Baker struck it rich and also the story of Alphonzo Bowers, who invented the first hydraulic gold dredge. **8.5" X 11", 54 ppgs, Retail Price: $9.99**

More Mining Books

Prospecting and Developing A Small Mine - Topics covered include the classification of varying ores, how to take a proper ore sample, the proper reduction of ore samples, alluvial sampling, how to understand geology as it is applied to prospecting and mining, prospecting procedures, methods of ore treatment, the application of drilling and blasting in a small mine and other topics that the small scale miner will find of benefit. **8.5" X 11", 112 ppgs, Retail Price: $11.99**

Timbering For Small Underground Mines - Topics covered include the selection of caps and posts, the treatment of mine timbers, how to install mine timbers, repairing damaged timbers, use of drift supports, headboards, squeeze sets, ore chute construction, mine cribbing, square set timbering methods, the use of steel and concrete sets and other topics that the small underground miner will find of benefit. This volume also includes twenty eight illustrations depicting the proper construction of mine timbering and support systems that greatly enhance the practical usability of the information contained in this small book. **8.5" X 11", 88 ppgs. Retail Price: $10.99**

Timbering and Mining - A classic mining publication on Hard Rock Mining by W.H. Storms. Unavailable since 1909, this rare publication provides an in depth look at American methods of underground mine timbering and mining methods. Topics include the selection and preservation of mine timbers, drifting and drift sets, driving in running ground, structural steel in mine workings, timbering drifts in gravel mines, timbering methods for driving shafts, positioning drill holes in shafts, timbering stations at shafts, drainage, mining large ore bodies by means of open cuts or by the "Glory Hole" system, stoping out ore in flat or low lying veins, use of the "Caving System", stoping in swelling ground, how to stope out large ore bodies, Square Set timbering on the Comstock and its modifications by California miners, the construction of ore chutes, stoping ore bodies by use of the "Block System", how to work dangerous ground, information on the "Delprat System" of stoping without mine timbers, construction and use of headframes and much more. This volume provides a reference into not only practical methods of mining and timbering that may be employed in narrow vein mining by small miners today, but also rare insights into how mines were being worked at the turn of the 19th Century. **8.5" X 11", 288 ppgs. Retail Price: $24.99**

A Study of Ore Deposits For The Practical Miner - Mining historian Kerby Jackson introduces us to a classic mining publication on ore deposits by J.P. Wallace. First published in 1908, it has been unavailable for over a century. Included are important insights into the properties of minerals and their identification, on the occurrence and origin of gold, on gold alloys, insights into gold bearing sulfides such as pyrites and arsenopyrites, on gold bearing vanadium, gold and silver tellurides, lead and mercury tellurides, on silver ores, platinum and iridium, mercury ores, copper ores, lead ores, zinc ores, iron ores, chromium ores, manganese ores, nickel ores, tin ores, tungsten ores and others. Also included are facts regarding rock forming minerals, their composition and occurrences, on igneous, sedimentary, metamorphic and intrusive rocks, as well as how they are geologically disturbed by dikes, flows and faults, as well as the effects of these geologic actions and why they are important to the miner. Written specifically with the common miner and prospector in mind, the book will help to unlock the earth's hidden wealth for you and is written in a simple and concise language that anyone can understand. **8.5" X 11", 366 ppgs. Retail Price: $24.99**

Mine Drainage - Unavailable since 1896, this rare publication provides an in depth look at American methods of underground mine drainage and mining pump systems. This volume provides a reference into not only practical methods of mining drainage that may be employed in narrow vein mining by small miners today, but also rare insights into how mines were being worked at the turn of the 19th Century. **8.5" X 11", 218 ppgs. Retail Price: $24.99**

Fire Assaying Gold, Silver and Lead Ores - Unavailable since 1907, this important publication was originally published by the Mining and Scientific Press and was designed to introduce miners and prospectors of gold, silver and lead to the art of fire assaying. Topics include the fire assaying of ores and products containing gold, silver and lead; the sampling and preparation of ore for an assay; care of the assay office, assay furnaces; crucibles and scorifiers; assay balances; metallic ores; scorification assays; cupelling; parting' crucible assays, the roasting of ores and more. This classic provides a time honored method of assaying put forward in a clear, concise and easy to understand language that will make it a benefit to even beginners. **8.5" X 11", 96 ppgs. Retail Price: $11.99**

Methods of Mine Timbering - Originally published in 1896, this important publication on mining engineering has not been available for nearly a century. Included are rare insights into historical methods of timbering structural support that were used in underground metal mines during the California that still have a practical application for the small scale hardrock miner of today. **8.5" X 11", 94 ppgs. Retail Price: $10.99**

The Enrichment of Copper Sulfide Ores - First published in 1913, it has been unavailable for over a century. Topics include the definition and types of ore enrichment, the oxidation of copper ores, the precipitation of metallic sulfides. Also included are the results of dozens of lab experiments pertaining to the enrichment of sulfide ores that will be of interest to the practical hard rock mine operator in his efforts to release the metallic bounty from his mine's ore. **8.5" X 11", 92 ppgs. Retail Price: $9.99**

A Study of Magmatic Sulfide Ores - Unavailable since 1914, this rare publication provides an in depth look at magmatic sulfide ores. Some of the topics included are the definition and classification of magmatic ores, descriptions of some magmatic sulfide ore deposits known at the time of publication including copper and nickel bearing pyrrohitic ore bodies, chalcopyrite-bornite deposits, pyritic deposits, magnetite-ileminite deposits, chromite deposits and magmatic iron ore deposits. Also included are details on how to recognize these types of ore deposits while prospecting for valuable hardrock minerals. **8.5" X 11", 138 ppgs. Retail Price: $11.99**

The Cyanide Process of Gold Recovery - Unavailable since 1894 and released under the name "The Cyanide Process: Its Practical Application and Economical Results", this rare publication provides an in depth look at the early use of cyanide leaching for gold recovery from hardrock mine ores. This volume provides a reference into the early development and use of cyanide leaching to recover gold. **8.5" X 11", 162 ppgs. Retail Price: $14.99**

California Gold Milling Practices - Unavailable since 1895 and released under the name "California Gold Practices", this rare publication provides an in depth look at early methods of milling used to reduce gold ores in California during the late 19th century. This volume provides a reference into the early development and use of milling equipment during the earliest years of the California Gold Rush up to the age of the Industrial Revolution. Much of the information still applies today and will be of use to small scale miners engaging in hardrock mining. **8.5" X 11", 104 ppgs. Retail Price: $10.99**

Leaching Gold and Silver Ores With The Plattner and Kiss Processes - Mining historian Kerby Jackson introduces us to a classic mining publication on the evaluation and examination of mines and prospects by C.H. Aaron. First published in 1881, it has been unavailable for over a century and sheds important light on the leaching of gold and silver ores with the Plattner and Kiss processes. **8.5" X 11", 204 ppgs. Retail Price: $15.99**

The Metallurgy of Lead and the Desilverization of Base Bullion - First published in 1896, it has been unavailable for over a century and sheds important light on the the recovery of silver from lead based ores. Some of the topics include the properties of lead and some of its compounds, lead ores such as galenite, anglesite, cerussite and others, the distribution of lead ores throughout the United States and the sampling and assaying of lead ores. Also covered is the metallurgical treatment of lead ores, as well as the desilverization of lead by the Pattinson Process and the Parkes Process. Hofman's text has long been considered one of the most important early works on the recovery of silver from lead based ores. 8.5" X 11", 452 ppgs. **Retail Price: $29.99**

Ore Sampling For Small Scale Miners - First published in 1916, it has been unavailable for over a century and sheds important light on historic methods of ore sampling in hardrock mines. Topics include how to take correct ore samples and the conditions that affect sampling, such as their subdivision and uniformity. Particular detail is given to methods of hand sampling ore bodies by grab sample, pipe sample and coning, as well as sampling by mechanical methods. Also given are insights into the screening, drying and grinding processes to achieve the most consistent sample results and much more. 8.5" X 11", 124 ppgs. **Retail Price: $12.99**

The Extraction of Silver, Copper and Tin from Ores - First published in 1896, it has been unavailable for over a century and sheds important light on how historic miners recovered silver, copper and tin from their mining operations. The book is split into three sections, including a discussion on the Lixiviation of Silver Ores, the mining and treatment of copper ores as practiced at Tharsis, Spain and the smelting of tin as it was practiced by metallurgists at Pulo Brani, Singapore. Also included is an overview and analysis of these historic metal recovery methods that will be of benefit to those interested in the extraction of silver, copper and tin from small mines. 8.5" X 11", 118 ppgs. **Retail Price: $14.99**

The Roasting of Gold and Silver Ores - First published in 1880, it has been unavailable for over a century and sheds important light on how historic miners recovered gold and silver rom their mining operations. Topics include details on the most important silver and free milling gold ores, methods of desulphurization of ores, methods of deoxidation, the chlorination of ores, methods and details on roasting gold and silver ores, notes on furnaces and more. Also included are details on numerous methods of gold and silver recovery, including the Ottokar Hofman's Process, the Patera Process, Kiss Process, Augustin Process, Ziervogel Process and others. 8.5" X 11", 178 ppgs. **Retail Price: $19.99**

The Examination of Mines and Prospects - First published in 1912, it has been unavailable for over a century and sheds important light on how to examine and evaluate hardrock mines, prospects and lode mining claims. Sections include Mining Examinations, Structural Geology, Structural Features of Ore Deposits, Primary Ores and their Distribution, Types of Primary Ore Deposits, Primary Ore Shoots, The Primary Alteration of Wall Rocks, Alterations by Surface Agencies, Residual Ores and their Distribution, Secondary Ores and Ore Shoots and Vein Outcrops. This hard to find information is a must for those who are interested in owning a mine or who already own a lode mining claim and wish to succeed at quartz mining. 8.5" X 11", 250 ppgs. **Retail Price: $19.99**

Garnets: Their Mining, Milling and Utilization - First published in 1925, it has been unavailable since those days and sheds important light on the mining, milling and utilization of garnets. Included are details on the characteristics of garnets, where they are found and how they were mined. 78 ppgs, 10.99

Gemstones and Precious Stones of North America - Leading mining historian Kerby Jackson introduces us to a classic mining publication on the gems and precious stones of the United States, Canada and mexico. First published in 1890, it has been unavailable since those days and sheds important light on the gems and precious stones that may be found in North America. Included are chapters on diamonds, corundum, sapphire, ruby, topaz, emerald, disapore, spinel, turquoise, tourmaline, garnets, beyrl, peridot, zircon, quartz crystals, feldspars, pearls and many others. Included are details on where these gems and precious stones may be found throughout North America, as well as their characteristics. 360 ppgs, 24.99

Mining Camps and Mining Districts - First released in 1885 by Charles Howard Shinn under the title "Mining Camps: A Study in American Frontier Government", this publication offers a unique look at how early gold miners established their own forms of representative government during the California Gold Rush. Drawing on the the early mining codes of mideviel German miners in the Harz Mountains, on the mining customs of the Cornish tin miners and early Spanish mining laws introduced into California, the miners established the first governments in the American West. 340 ppgs, 24.99

BLM Field Handbook for Mineral Examiners - Leading mining historian Kerby Jackson introduces us to a classic mining publication on mine evaluation. First published in 1962, this work sheds important light on the techniques of BLM Mineral Examiners to perform validity on mining claims. 132 ppgs, 10.99

<u>**Six Months In The Gold Mines During The California Gold Rush**</u> - Unavailable since 1850, this important work is a first hand account of one "49'ers" personal experience during the great California Gold Rush, shedding important light on one of the most exciting periods in the history of not only California, but also the world. Compiled from journals written between 1847 and 1849 by E. Gould Buffum, a native of New York, "Six Months In The Gold Mines During The California Gold Rush" offers a rare look into the day to day lives of the people who came to California to work in her gold mines when the state was still a great frontier. **8.5" X 11", 290 ppgs. Retail Price: $19.99**

<u>**The Discovery of Gold in Australia**</u> - First published in 1852, it has been unavailable since those days and sheds important light on Australia's gold mining history. Included are rare communications between British agents and the British Crown when gold was first discovered in Australia in 1851. This rare text contains hard to find details on Australia's first mining camps and Britain's early attempts to provide for the orderly regulation of gold mines in that part of the world. Also of interest are hard to find extracts of articles that appeared in the early colonial newspapers that did their best to report on Australia's gold rush as it took place.
102 ppgs, 10.99